W9-CJM-875

CLINICAL BIOCHEMISTRY MADE RIDICULOUSLY SIMPLE

Stephen Goldberg, M.D.
Associate Professor
Department of Anatomy and Cell Biology
Department of Family Medicine
University of Miami School of Medicine
Miami, Florida

MedMaster, Inc., Miami

Copyright © 1988, 1991 by MedMaster, Inc.
Second Printing, 1988 *Third Printing*, 1991

All rights reserved. This book is protected by copyright. No part of it may be reproduced, stored in a retrieval system, or transmitted in any form or by any means, electronic, mechanical, photocopying, recording or otherwise, without written permission from the copyright owner.

ISBN #0-940780-10-0

Made in the United States of America

Published by
MedMaster, Inc.
P.O. Box 640028
Miami, FL 33164

TO MY PARENTS, HARRY AND DINAH GOLDBERG, WHO HAVE PROVIDED LOVE, GUIDANCE, AND OPPORTUNITY.

PREFACE

This book focuses on clinically relevant biochemistry, for medical students and other health professionals.

There is a great difference between the research-oriented needs of the biochemistry graduate student and the clinical needs of the medical student. A book for graduate students needs to emphasize research methods and functionally important points. A book for medical students needs to provide the basic conceptual background that will allow the student to understand disease mechanisms, clinical laboratory tests, and drug effects.

The first step in preparing this book was the selection of that biochemical information with the greatest clinical relevance. The second step was an attempt to present that information in a way that optimally facilitates learning and retention.

I have tried to present an overall conceptual picture rather than focusing on fine detail. Courses frequently deliver an overwhelming amount of esoteria with the expectation that the student will eventually integrate this into an overall view. Commonly, the overall view never gels and the student is left with isolated points that have little apparent linkage and are quickly forgotten after the exam. This book attempts at the **outset** to present the major chemical reactions in one central map of Biochemistryland that may be conceptualized quickly and is the central focus of the book. After providing the overall view, the text centers on more detailed clinical material. I have tried to use visual imagery, humor, and other memory techniques, not in disrespect for the field, but as educational methods that should be used more often in medical education.

The boundaries between biochemistry, physiology, pharmacology, microbiology, and immunology are fuzzy, but it is necessary to draw the line somewhere. A number of topics, therefore, such as electrolyte and acid-base balance have not been included as they overlap with physiology courses and don't quite fit into the Biochemistryland map. Certain points in pharmacology and microbiology are presented where they pertain directly to the chemical reactions at hand, but these are presented briefly. The Appendices review isomer and enzyme terminology.

I have not emphasized structure in this book, as it is the flow of events, rather than specific structural formulae that are most relevant from a clinical standpoint. I have, however, included a structural index for reference to specific structures when appropriate.

This book is not intended as a replacement for standard reference texts, but is a supplement to show the overall picture, in conjunction with the reference text. It should also be useful for Board review. The idea is that students learn better when they have two kinds of books—a small book that directly shows the overall view, and a basic reference text.

I thank Dr. Frans Huijing for his many helpful suggestions in all areas of this book. Dr. Bert Pressman provided a number of useful discussions. Charles Messing drew the Biochemistryland map, the cover illustration, and the text figures. I am grateful to the Boehringer-Mannheim Biochemical Co. for their kind permission to reproduce, in the Structural Index, many of the structural formulae from their excellent map of the pathways in biochemical metabolism.

CONTENTS

CHAPTER 1. AN OVERVIEW OF BIOCHEMISTRYLAND

Biochemistryland is a biochemical amusement park. As seen from high in the air in figure 1.1, Biochemistryland consists of a number of sections: THE MAIN POWER HOUSE, CARBOHYDRATELAND, LIPIDLAND, THE AMINO ACID MIDWAY, THE DNA FUNHOUSE, PORPHY'S HEMELAND, COMBO CIRCLE, and an INFIRMARY. A closer look (fig. 1.2) reveals finer details, such as interconnections between the sections and subdivisions of the sections themselves.

1. **THE MAIN POWERHOUSE** is the **key** energy source of Biochemistryland and, in fact, is **key-shaped.** Its Main Hallway (fig. 1.2) leads to a ferris wheel (Krebs cycle) run by a powerful Generator (oxidative phosphorylation). There is an Energy Hall of Fame (HOF). The Main Powerhouse is situated near a Saloon (alcohol metabolism).
2. **CARBOHYDRATELAND** is shaped like a slice of cake. It contains a small, accessory powerhouse (the Penthouse Powerhouse) (fig. 1.2), a carbohydrate Storage Room, an Ice Cream Parlor, and a Conjugation area.
3. **LIPIDLAND** is shaped like a string of three hot dogs. It contains, in its western sector, Phosphatidylywink Village (fig. 1.2), a Frog Pond, Sphingo's Curio Shop, and a Lipid Storage Room. In Mid-Lipidland, there are two Roller Coasters (an up and a down roller coaster, for fatty acid biosynthesis and degradation), and a musical review by the Ketones in the Ketone Playhouse. In the East, progressing through Channel #5, one arrives at the dark and mysterious Sterol Forest, where one may search for sex, but must beware of falling into Bile Bog.
4. **THE AMINO ACID MIDWAY** is a broad field that stretches throughout Biochemistryland and connects with many areas. It has a (Urea) Restroom (fig. 1.2).
5. **THE DNA FUNHOUSE** contains a sideshow with mutants and other amazing transformations. It also has a (Uric Acid) Restroom (fig. 1.2).
6. **PORPHY'S HEMELAND** is for people who like real blood and guts adventure. Buy your pinwheels here.
7. **COMBO CIRCLE** is a triple theatre that stars the glycoproteins, glycolipids, and lipoproteins (fig. 1.2).
8. The **INFIRMARY** is stocked with drugs, including vitamins, hormones, and minerals. An infirmary is necessary, as Biochemistryland is a hazardous place where one may find diseases of diverse nature, which are indicated by red-encircled numbers of the main map. The main map may be found on the inside of the back cover of the book. It shows Biochemistryland in finer detail than figures 1.1 and 1.2. Discussion of the diseases on the map may be found in the Clinical Review (chapter 10).

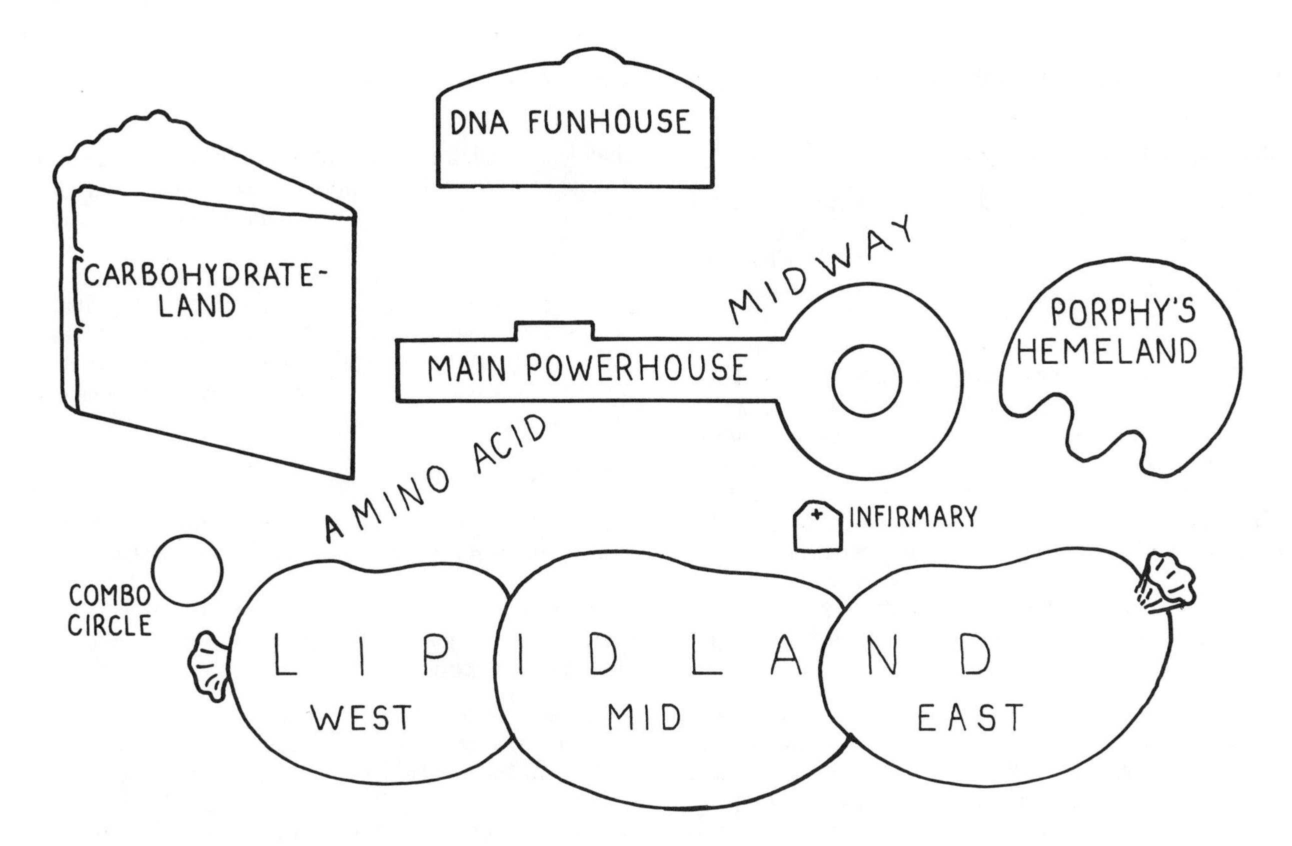

Fig. 1.1. An aerial view of Biochemistryland.

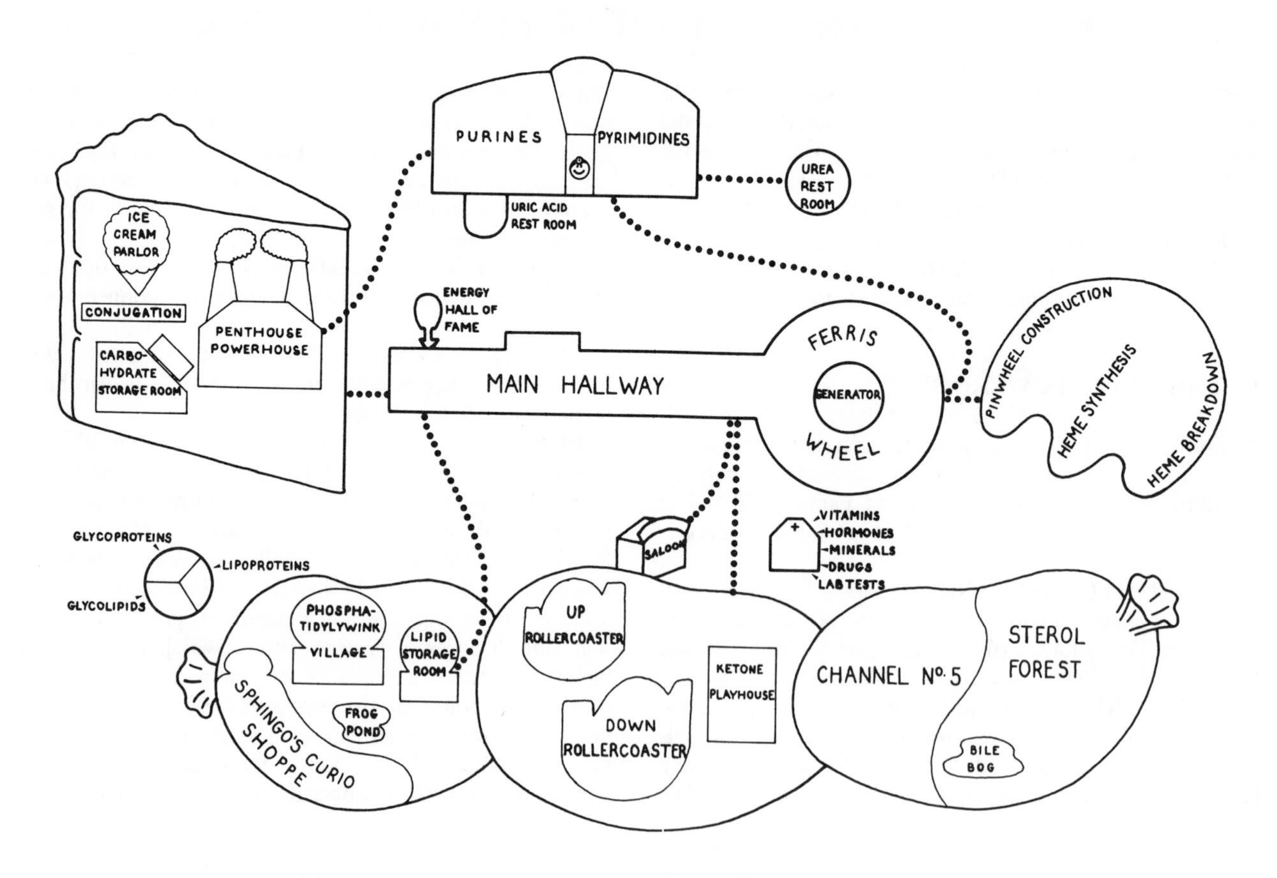

Fig. 1.2. A closer look at Biochemistryland. HOF, Energy Hall of Fame.

Where is Biochemistryland?

The Biochemistryland map is a way of viewing the key biochemical reactions of the body. The map, though, does not correspond to any known human anatomy. Why, then, should one bother to place the various biochemical reactions in such an artificial format? Why not simply draw a liver, intestine, muscle, brain, etc. and indicate the various biochemical reactions therein so that one may know where in the body the individual reactions occur? The problem with this approach is that a particular chemical reaction often occurs in many organ systems. If the individual organs were drawn, with the idea of including their chemical reactions inside them, there would be extensive duplication of pathways and an uninterpretable map. If one wishes to draw the individual reactions only once, one needs a different format. The format of the Biochemistryland map allows this, in a way that can be appreciated visually as a whole.

Although organs are not shown on the map, it is nonetheless important for the clinician to know which reactions correspond to different organ systems. Certain reaction steps are almost universal throughout the body (e.g., glycolysis). Other reactions are more confined to particular organs (e.g. thyroxine production). When a chemical reaction is organ-specific, a laboratory test that detects an abnormal amount of the particular chemical in question may alert the clinician to the anatomical location of the specific pathology. For instance, elevation of serum SGPT (ALT) enzyme levels may provide a clue as to the presence of liver injury. In view of the clinical importance of knowing the anatomical localizations of particular reactions, each chapter includes at the end a discussion of how the particular Biochemistryland zone corresponds to locations within the body.

How To Use This Book

Keep the main map (inside of back cover) in front of you, as the map is the central focus for all the chapters. If you like, the Structural Index at the end of the book may be used to check specific biochemical structures. Structure, though, is not as important clinically as is the general flow of reactions.

Diseases indicated on the map by the red-encircled numbers are discussed in the chapter entitled CLINICAL REVIEW. The numbers within the red circles correspond to the numbers in the CLINICAL REVIEW chapter. For instance, 36, encircled in red on the map, refers to clinical condition #36 as discussed in the CLINICAL RE-

VIEW under #36. You may wish to wait until reading all the chapters before studying these diseases as a whole. Alternatively, you may wish to examine them as you read each chapter to see the clinical relevance of the points under discussion. Vitamin-containing structures are indicated in green on the main map.

Throughout the book, map locations are given to direct the reader to specific points on the map. For instance "G-4" refers to coordinates "G-4" on the map. The terms "-ate" and "ic acid" (as in pyruvate and pyruvic acid) are used interchangeably, the "-ate" form being the ionized ($-COO^-$) form of the acid ($-COOH$). Appendix I reviews the terms used in describing isomers. Appendix II defines the terminology in enzyme classification.

Fix in mind figures 1.1 and 1.2 before proceeding.

CHAPTER 2. THE MAIN POWERHOUSE

Biochemistryland needs energy to keep running. There are a number of sources of this fuel. The ultimate source of energy is the sun, which enables plants to make glucose from CO_2 and H_2O during photosynthesis. Biochemistryland doesn't burn glucose in the strict sense of a fire. Fire might be good enough to make steam and drive a steam engine. The body, however, would burn up if it crudely used heat to run. Instead, the energy released during the breakdown of glucose and other molecules to CO_2 and H_2O is released gradually to form energy-containing packets, the most important of which is the ATP (adenosine triphosphate) molecule. ATP is then used as an energy source in driving many biochemical reactions. It is estimated that we use more than half our body weight each day in ATP.

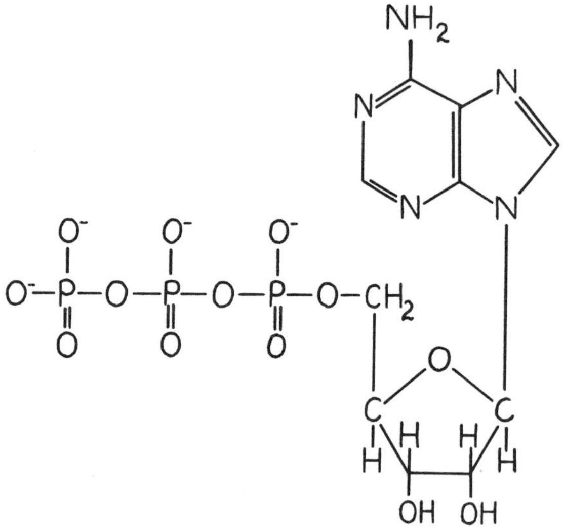

ATP

ATP, while a good energy packet, is not a good fuel storage molecule, as it is used quickly after being formed. Better storage forms of energy are glycogen (in the Carbohydrateland Storage area) and triglycerides (in the Lipidland Storage area). When necessary, these storage molecules can be broken down and used to regenerate ATP.

Let us start our tour of the Main Powerhouse with the Energy Hall Of Fame.

The Energy Hall Of Fame

If ATP were the only kind of "energy molecule", the Energy Hall Of Fame would be small, indeed. The fact is that there are many kinds of molecules capable of supplying energy besides ATP. In each case, ATP included, the molecule has the additional talent of doing something else, during the process of delivering energy, that contributes to the particular chemical reaction at hand. In ATP's case it is the transfer of a phosphate group, a common occurrence throughout Biochemistryland. For other molecules, it may be the transfer of single or double carbon groups, hydrogen atoms, electrons, or other things, as shown in figure 2.1 Nor are these listed ones the only high energy molecules in the body. The ones listed in the Energy Hall of Fame were placed there because they have general functions throughout Biochemistryland. Other molecules may contribute significant energy but act at only one spot on the map. For instance, PEP is a very high energy molecule that supplies phosphate groups to ADP (adenosine diphosphate) to form ATP (it is full of "PEP") in the reaction:

PEP + ADP ⟶ PYRUVATE + ATP
(D-7)

Reducing agents are those that supply hydrogen atoms, or electrons in chemical reactions. (**Oxidizing** agents receive hydrogen atoms or electrons). Molecules such as NADH, NADPH, and $FADH_2$ are reducing agents that release energy when reacting and are said to have "reducing power". In general, biosynthesized molecules become more reduced with synthesis, and reductive biosynthesis, especially through the use of molecules like NADPH, are important in Biochemistryland. NADPH differs from NADH in that the energy supplied by NADPH generally is used for a variety of biosyntheses rather than the generation of ATP. We will encounter the uses of NADH and $FADH_2$ in ATP production when we explore the Main Powerhouse Ferris Wheel Generator (oxidative phosphorylation).

Chemical reactions procede in the direction in which the bonds are more stable and in which energy is released. Such energy may be dissipated as heat or captured and used for other things, like making ATP molecules. It is an important principle that when several reactions occur concurrently the final result depends on the **net** energy loss. Thus:

A ⟷ B + C	△ +6 kcal/mol (Reaction goes to A as it would require energy input to go to B + C)
B ⟷ D	△ − 9 kcal/mol (Reaction goes to D with net loss of energy)
A ⟷ C + D	△ − 3 kcal/mol (Net reaction is to the right, with net loss of energy)

The net reaction in the above series is to the right (A goes

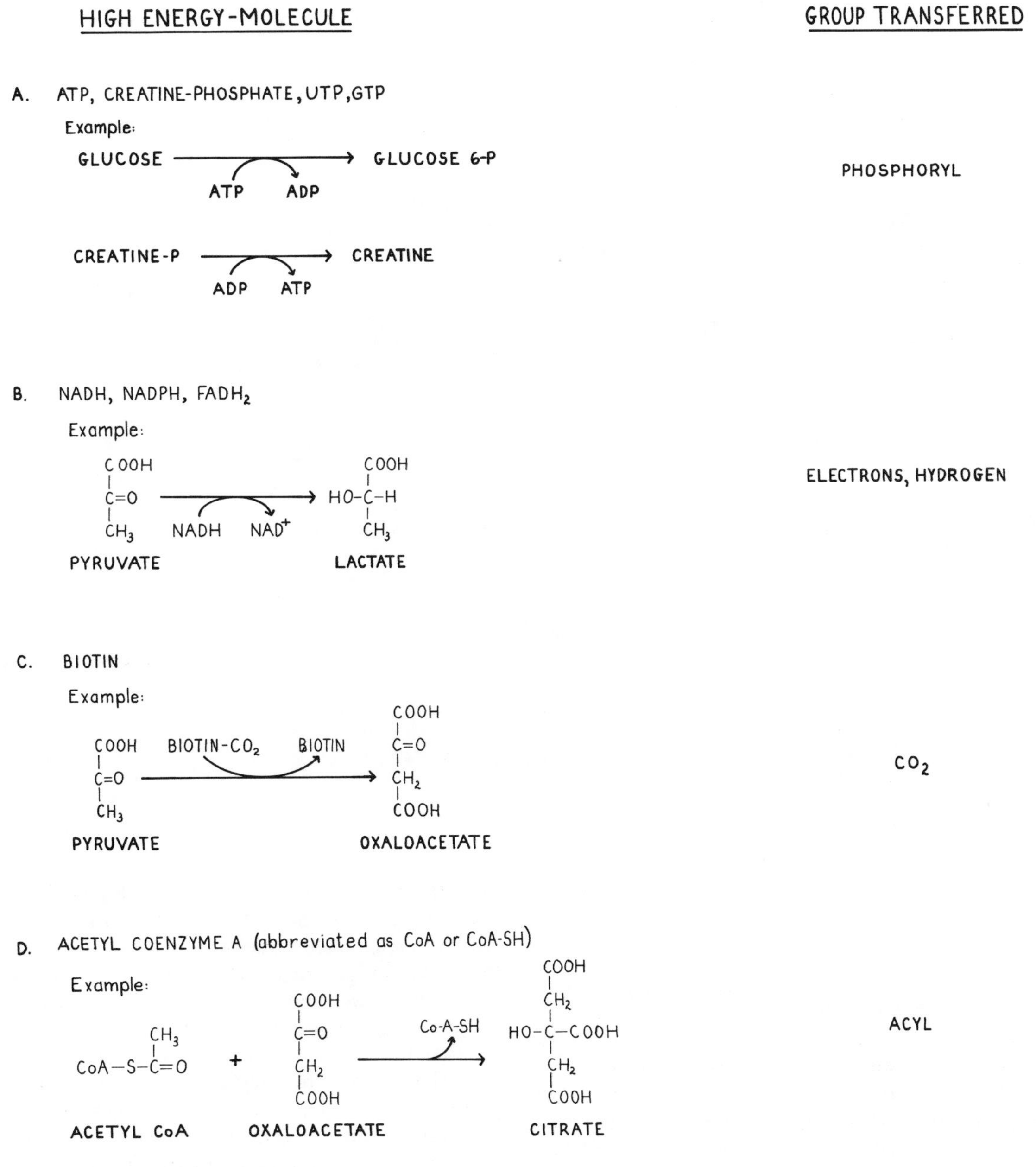

HIGH ENERGY-MOLECULE	GROUP TRANSFERRED
A. ATP, CREATINE-PHOSPHATE, UTP, GTP	
Example: GLUCOSE → GLUCOSE 6-P (ATP → ADP)	PHOSPHORYL
CREATINE-P → CREATINE (ADP → ATP)	
B. NADH, NADPH, $FADH_2$	
Example: PYRUVATE (COOH–C=O–CH_3) → LACTATE (COOH–HO-C-H–CH_3) (NADH → NAD^+)	ELECTRONS, HYDROGEN
C. BIOTIN	
Example: PYRUVATE (COOH–C=O–CH_3) → OXALOACETATE (COOH–C=O–CH_2–COOH) (BIOTIN-CO_2 → BIOTIN)	CO_2
D. ACETYL COENZYME A (abbreviated as CoA or CoA-SH)	
Example: ACETYL CoA (CoA–S–C(=O)–CH_3) + OXALOACETATE (COOH–C=O–CH_2–COOH) → CITRATE (COOH–CH_2–HO-C-COOH–CH_2–COOH) (→ Co-A-SH)	ACYL

Fig. 2.1. Examples from The Energy Hall of Fame. These molecules not only deliver energy, but transfer special groups in the process. Acyl, fatty acid-bearing; ADP, adenosine diphosphate; ATP, adenosine triphosphate; dUMP deoxyuridine monophosphate; FAD, flavin adenine dinucleotide; GTP, guanosine triphosphate; NADH, nicotinamide adenine dinucleotide; NADP, nicotinamide adenine dinucleotide phosphate; P, phosphate; TMP, thymidine monophosphate; UDP, uridine diphosphate; UTP, uridine triphosphate.

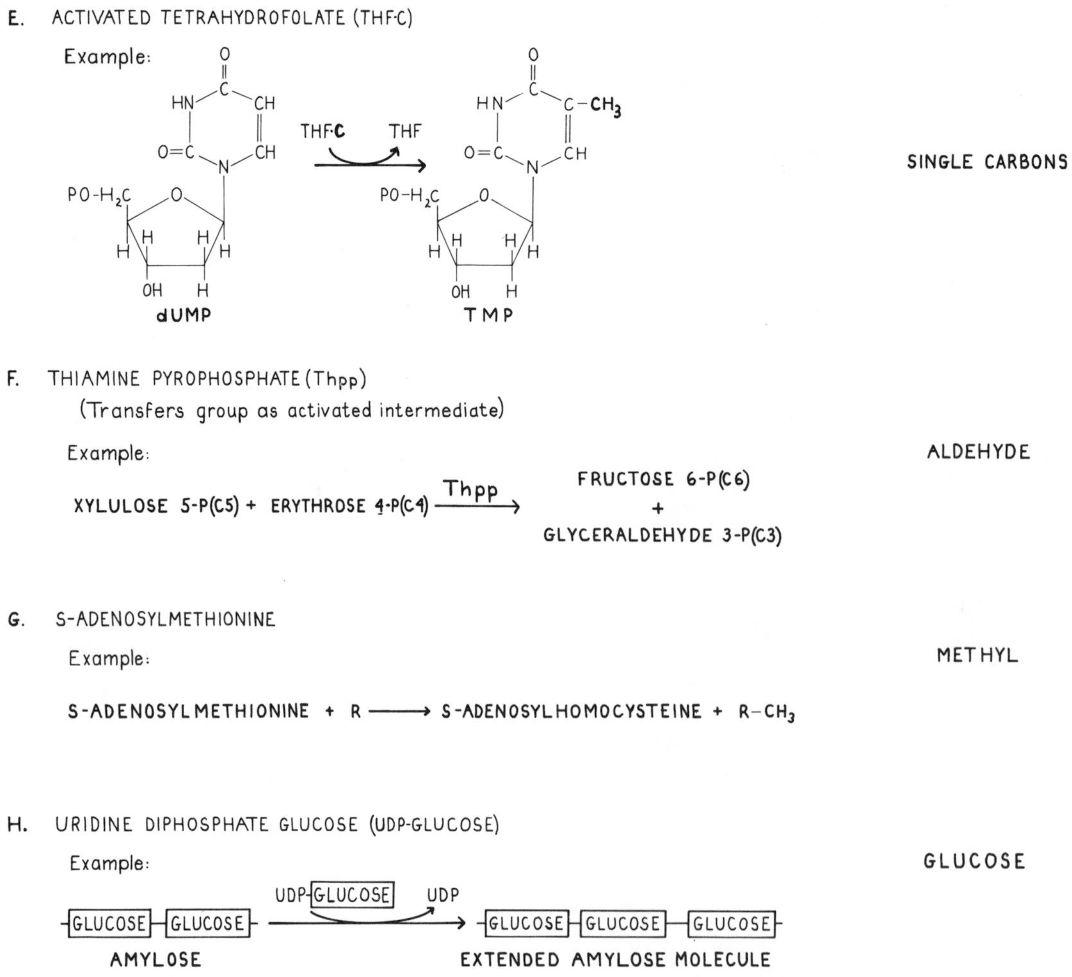

E. ACTIVATED TETRAHYDROFOLATE (THF·C)

Example:

dUMP —(THF·C → THF)→ TMP

SINGLE CARBONS

F. THIAMINE PYROPHOSPHATE (Thpp)
(Transfers group as activated intermediate)

Example:

XYLULOSE 5-P(C5) + ERYTHROSE 4-P(C4) —Thpp→ FRUCTOSE 6-P(C6) + GLYCERALDEHYDE 3-P(C3)

ALDEHYDE

G. S-ADENOSYLMETHIONINE

Example:

S-ADENOSYLMETHIONINE + R ⟶ S-ADENOSYLHOMOCYSTEINE + R–CH_3

METHYL

H. URIDINE DIPHOSPHATE GLUCOSE (UDP-GLUCOSE)

Example:

–GLUCOSE–GLUCOSE– —(UDP-GLUCOSE → UDP)→ –GLUCOSE–GLUCOSE–GLUCOSE–

AMYLOSE ⟶ EXTENDED AMYLOSE MOLECULE

GLUCOSE

Fig. 2–1 (Continued)

to C + D) even though the first reaction (A to B + C) has a tendency to move to the left.

In all reactions, the **first law of thermodynamics** applies: energy is conserved; the total energy on one side of an equation equals the total on the other. According to the **second law of thermodynamics,** things tend to procede from a state of order to a state of disorder, **entropy** being a term used to describe the degree of disorder. Thus, gas molecules locked up and concentrated in a container tend to diffuse away when the container is opened, rather than the reverse; heat flows from a hot object to a cold one rather than the opposite; chemical reactions prefer to go in the direction in which energy is released and dissipated. There has been much confusion about this point when considering the human body, because chemical reactions often procede from simple molecules to more complex ones; complex macroscopic structures develop from simpler ones; the body grows and is maintained rather than decays; the body appears to create greater order rather than greater disorder. The situation is not, however, in conflict with the second law of thermodynamics. Although chemical syntheses do procede partly in a way that produces greater order in body structure and function, they are accompanied by even greater disorder produced in the dissipation of energy in the course of these reactions. We don't see this energy dissipation but do see the orderly aspect of things. The net result, however, is greater disorder than order. Exactly how it is that the body came about developing mechanisms to produce such complex reactions is another matter. This would involve a speculative discussion about chemical evolution, however, and is beyond the province of this book (fortunately).

Chemical reactions may be catalyzed by special proteins called enzymes. Enzymes do not supply additional energy, or change the direction of reactions. The reaction continues in the same direction, with the same eventual results. What an enzyme does is to **speed up** a reaction that ordinarily might take a very long time to occur.

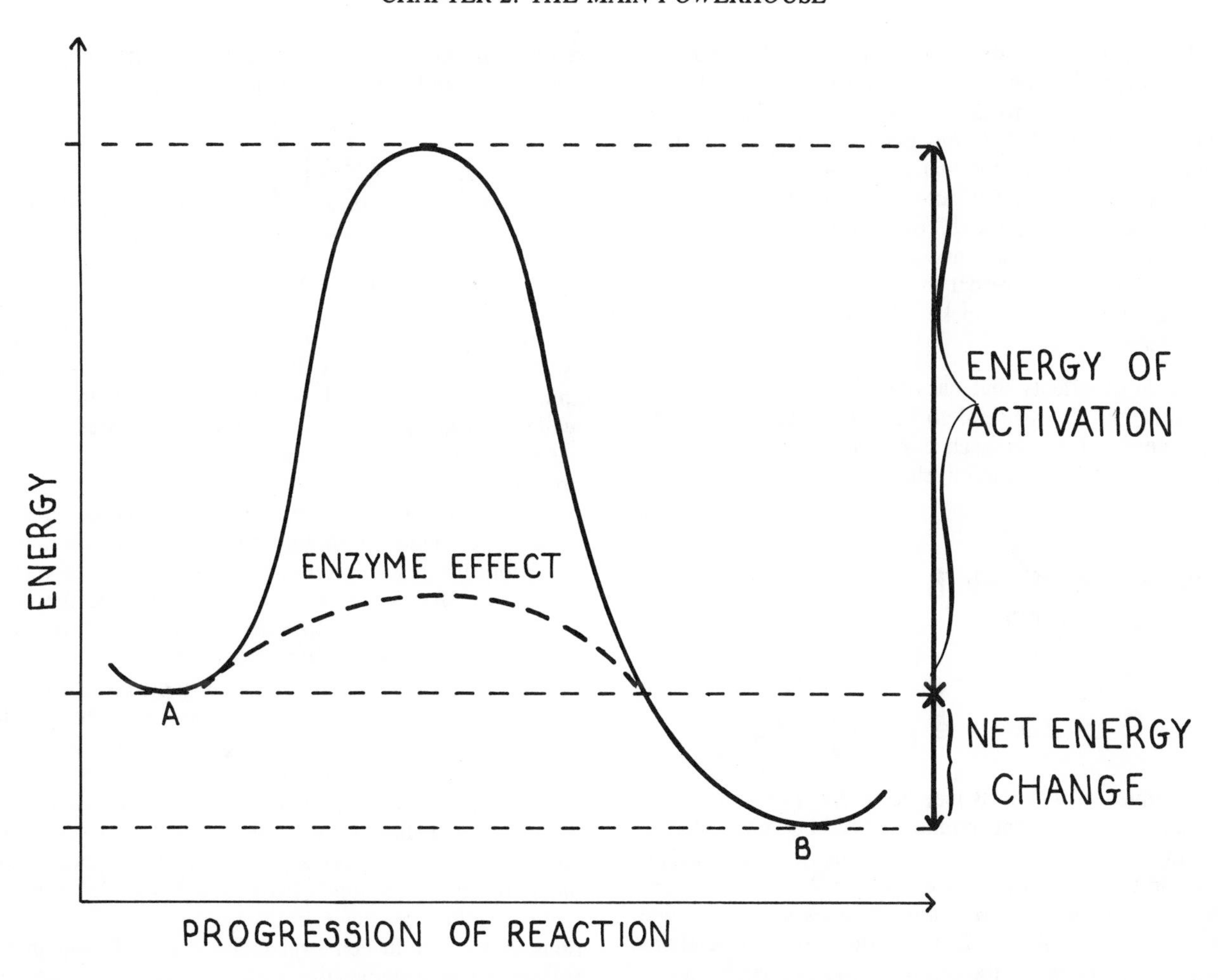

Fig. 2.2. The energy hill of activation. The reaction from A to B may take a long time if there is a high energy requirement to activate the reaction. Enzymes lower this energy hill of activation and speed up the reaction.

Normally, substrates in a given reaction need a certain level of energy (**energy of activation**) to react. At any given time, some molecules are randomly in a higher state of activation than others and will cross over the energy hill of activation. This process is facilitated by enzymes, which lower the energy hill of activation (fig. 2.2). Enzymes sometimes do this by facilitating the alignment of the reactants with one another so that the latter do not have to travel around and align with one another randomly.

The **direction** of a reaction depends on the direction of net energy change in the reaction and on the concentration of reactants. The **rate** of a reaction, however, does not depend on the net energy change. The rate may be quite slow even in a reaction in which the net energy change is great. Rate depends largely on the energy hill of activation, which is an entity apart from the net energy change (fig. 2.2). Increased temperature also increases the reaction rate, as molecules move faster and have an greater chance of contacting one another and a better chance to pass over the energy hill of activation. Rate also depends on the concentrations of substrate and product.

If enzymes do not change the direction of a reaction, it is at first puzzling to note that many reactions in Biochemistryland contain two-way arrows, with one enzyme seeming to direct the reaction one way and the other enzyme the other way. For instance:

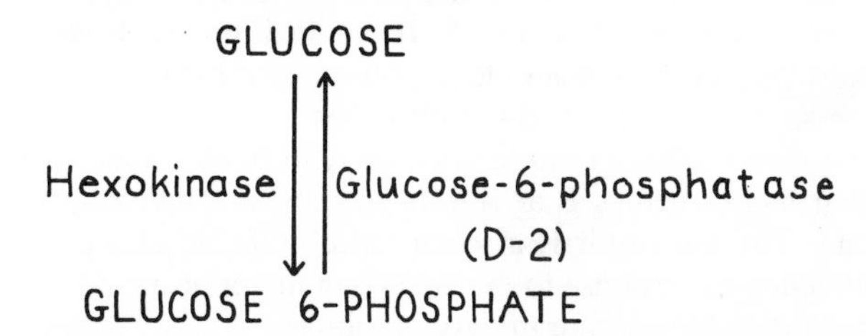

Hexokinase facilitates the reaction toward glucose 6-phosphate whereas glucose 6-phosphatase facilitates the reaction toward glucose. On examining each of these reactions more closely, however, note that the reactions each way are really **different,** containing different substrates, depending on the direction. Thus, one reaction is: glucose + ATP forms glucose 6-P + ADP. The other is: glucose 6-P + H_2O forms glucose and Pi (inorganic phosphate). The enzyme that acts on one set of ingredients does not necessarily act for the other set. The reaction that actually occurs will depend on the availability of the particular substrates and enzymes. Since glucose 6-phosphatase is not present in muscle cells, for instance, muscle cells do not make glucose from glucose 6-phosphate.

The large effects that enzymes have on reaction rate explains why it is useful to have two-way arrows with separate enzymes for each direction. For instance, consider the following reaction chain:

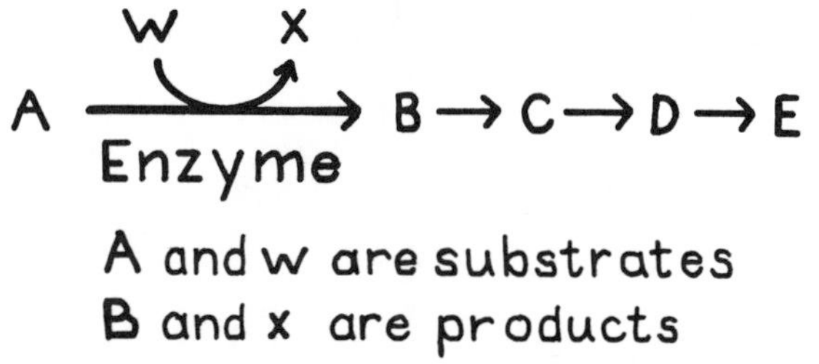

W X

A → B → C → D → E

Enzyme

A and W are substrates

B and X are products

The sequence of A to B may be reversed to some degree by an increased concentration of E, but only slightly. (Compare for instance, the difficulties one will experience in trying to reverse the reaction $Ag^{++} + Cl^{-} \longrightarrow AgCl_2$ (precipitate), a reaction that goes far to the right, by adding more $AgCl_2$. Reaction direction can be altered drastically, though if the reaction sequence is:

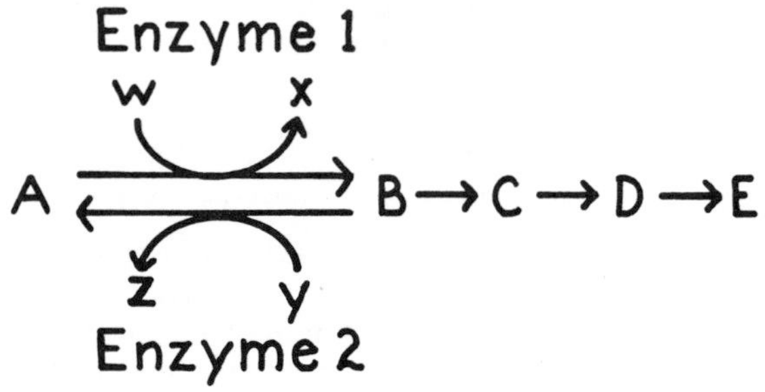

Enzyme 1

W X

A ⇄ B → C → D → E

Z Y

Enzyme 2

Enzyme 1 directs the reaction of A to B and enzyme 2 directs the reaction of B to A. If the final product E feeds back (negative feedback) to inhibit enzyme 1 (thus signaling an excess of E), and stimulates the action of enzyme 2 (positive feedback) the reaction of A to B will come to a near halt, whereas B to A will procede at a much faster rate. The net result is a much more dramatic change in direction in response to excess E than might occur if concentration alone were the driving force.

In general, it is important that enzyme reactions which procede in one direction do not occur simultaneously in the same location as those that procede in the opposite direction. Otherwise, the reactions will compete with one another, generating a continuous cycling that only wastes energy. For instance, consider:

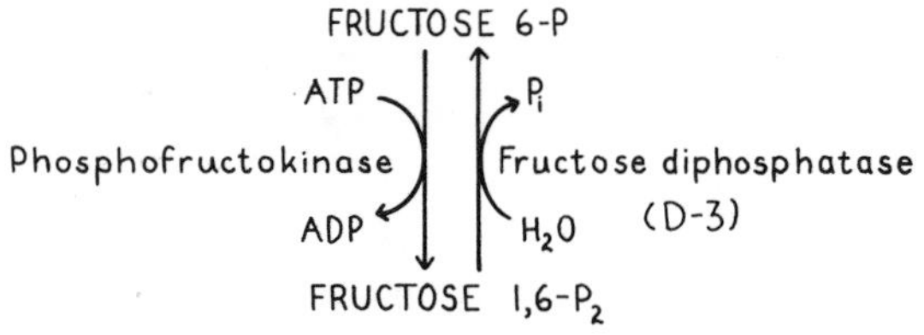

FRUCTOSE 6-P

ATP → ADP (Phosphofructokinase) | H_2O → P_i (Fructose diphosphatase (D-3))

FRUCTOSE 1,6-P_2

When the two-way reactions act together, as a continuous cycle, this wastes ATP, with the generation of heat. Some animals use this to their advantage for heat production, while hibernating. In humans pathological cases may occur in which two-way reactions such as these may inappropriately occur simultaneously, with excess heat production to the degree that may cause death. This is believed to occur in **malignant hyperthermia,** a rare condition that results from a peculiar reaction in certain individuals who are exposed to halothane anesthesia or certain other chemicals. In order to avoid the situation of competition within two-way reactions, the body commonly uses the strategy of negative (and positive) feedback to insure that both enzymes are not acting concurrently. Alternatively, it is common that separate direction reactions, while conveniently placed near one another on paper, are separated in the body, whether in different organelles, or in different organs. Commonly, biosynthetic reactions are separated within the body from biodegradation reactions. Figures 2.3 and 2.4 summarize, for future reference in the book, the different reactions that occur in cell organelles (fig. 2.3) and in the various organ systems (fig. 2.4).

The effects of enzymes should be distinguished from the effects of hormones. Enzymes play a direct role in controlling reaction rates. Hormones appear to act by directly or indirectly affecting the degree of enzyme synthesis or activation.

Let us now take a closer look at the key sections of the Main Powerhouse—the Main Hallway (glycolysis and gluconeogenesis), and the Ferris Wheel (Krebs cycle).

The Main Hallway and the Ferris Wheel

A primary conceptual step in the functioning of the Main Powerhouse is the splitting of glucose, a 6-carbon sugar, into two 3-carbon halves. This occurs right at the entrance to the Main Hallway, where glyceraldehyde 3-phosphate is formed. This in turn can eventually lose a carbon to form acetyl CoA, the key fuel of the Main Powerhouse ferris wheel. Imagine a ferris wheel in which two passengers get on, sit down, and then get off. There is no net change in the number of seats on the ferris wheel. Similarly, for every 2 carbons that get on the Krebs cycle, as acetyl groups from acetyl CoA, two carbons get off as CO_2, and there is no net change in the ingredients of the ferris wheel. (Actually the two carbons

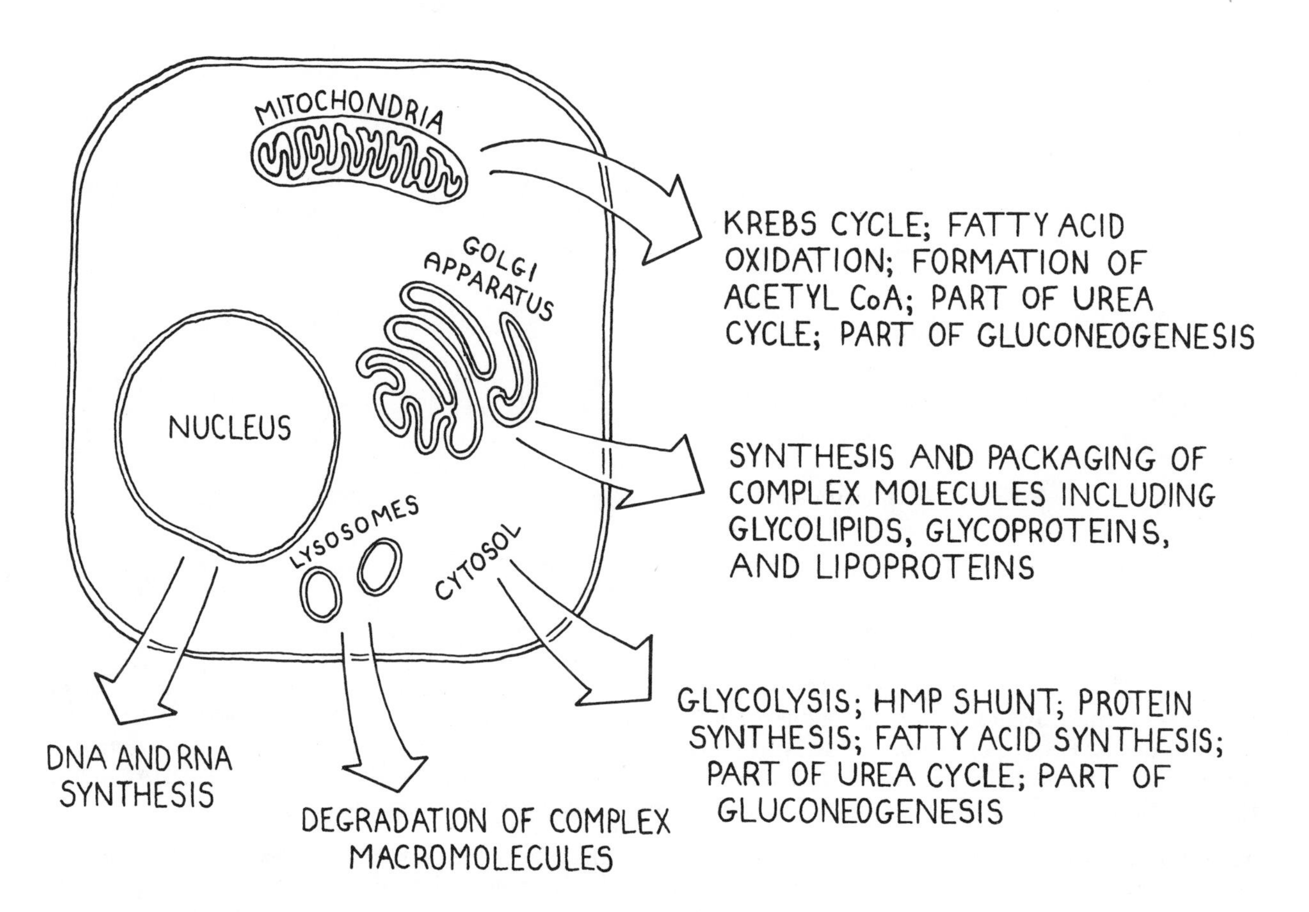

Fig. 2.3. The intracellular localization of some key biochemical reactions.

HOUSEKEEPING (GENERALIZED) ORGAN FUNCTIONS

Fatty acid oxidation (*)(**)
Glycolysis
Glycogen synthesis and breakdown
Krebs cycle and oxidative phosphorylation (**)
Protein synthesis (**)

(*) except in brain
(**) except in mature red blood cells

SPECIALIZED ORGAN FUNCTIONS	MAJOR SITES
Fatty acid synthesis	liver, fat cells
Gluconeogenesis	liver, kidney
Heme synthesis	bone marrow
HMP shunt	liver, fat cells, adrenal cortex mammary gland
Amino acid synthesis and breakdown	liver
Urea synthesis	liver
Cholesterol synthesis	liver
Steroid hormone synthesis	adrenal cortex, gonads

Fig. 2.4. The organ localization of some key biochemical reactions.

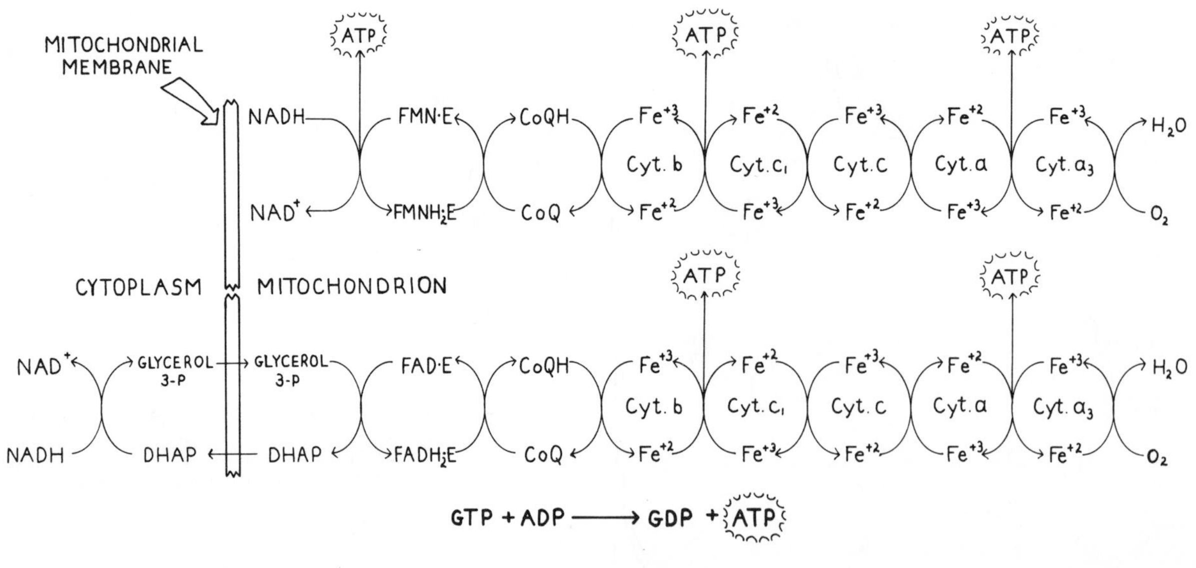

Fig. 2.5. The ferris wheel generator: oxidative phosphorylation. Intramitochondrial NADH yields 3 ATPs. $FADH_2$ yields 2 ATPs. Extramitochondrial NADH (generated in the cytosol in the Main Hallway) yields only 2 ATPs. This is a consequence of the fact that cytosol NADH cannot pass through the mitochondrial membrane. It must do so via Glycerol-3-P (the Glycerol Phosphate Shuttle), but in doing so, loses the potential of producing an ATP. One GTP produces one ATP. Cyt., cytochrome. The cytochromes are heme-containing proteins that participate in electron transport through valence changes in the heme iron. FMN, flavin mononucleotide; CoQH, coenzyme Q; DHAP, dihydroxyacetone phosphate; FAD, flavin adenine dinucleotide; NAD, nicotinamide adenine dinucleotide.

that get on are not the same two carbons that get off, but the principle is the same). The degradation of the two carbons to form CO_2 releases energy that is captured to form ATP. A small amount of ATP is formed in the Main Hallway (glycolysis), but most is produced at the level of the Ferris Wheel (Krebs cycle) and its generator (oxidative phosphorylation (fig. 2.5).

The acetyl CoA that gets on the ferris wheel can be continually replenished through glucose breakdown, or, mainly, through fatty acid degradation (oxidation), or by transformation of certain amino acids. What, however, produces the seats of the ferris wheel, or replenishes them when necessary? The seats cannot be replaced by acetyl CoA, which is merely a passenger. The chemicals of the ferris wheel can be restored in part by certain amino acids that can convert to Krebs cycle intermediates. There also is an important side step in which pyruvate can be directly converted to oxaloacetate (D-8).

A glance at the Biochemistryland map shows that fructose is another 6-carbon sugar that can split in two and be used as fuel to feed the Main Powerhouse.

As mentioned, most of the ATP is produced during reactions in the Krebs cycle, specifically during oxidative phosphorylation. However, the Krebs cycle requires O_2 to run. In the absence of O_2, **anaerobic** glycolysis can still occur (in the Main Hallway), with the production of some ATP. In order for this to happen though, the NADH produced in the step between glyceraldehyde 3-P and 1,3P_2-glycerate needs to be changed back to NAD^+ for reuse in the latter step. Normally NAD^+ restoration occurs in oxidative phosphorylation, which requires O_2. Without O_2, pyruvate instead changes to lactate, a step in which NAD^+ can be replaced. Consequently, there may be a buildup of blood lactate during vigorous exercise.

Although ATP is spun off during the Krebs cycle, this is not apparent on just looking at the ferris wheel. The ATP does not just come off directly from the ferris wheel. Rather, NADH, $FADH_2$, and GTP molecules, which do come off the ferris wheel are used to produce ATP in the Ferris Wheel generator (fig. 2.5). NADH and $FADH_2$ molecules supply electrons that are passed along an elaborate bucket brigade of molecules that ends in a reaction with O_2 and the release of energy that changes ADP to ATP. The term **oxidative phosphorylation** refers to the events in the Ferris Wheel generator: **Oxidation** by a series of reactions requiring O_2 at the end; **phosphorylation** of ADP to ATP in the same chain. Oxidation and phosphorylation are said to be **coupled** to one another. I.e., oxidation occurs concurrently with ATP generation.

Oxidation is stimulated by the presence of ADP. In the absence of ADP, oxidation slows. This provides a control mechanism wherein the rate of oxidation matches the need for ATP. Oxidation thus increases when ATP is low (and ADP high) and slows when ATP is high (and ADP low.)

It is possible to add uncoupling agents that prevent the formation of ATP. In such cases there is increased oxygen consumption and generation of heat. Hibernating animals may use this mechanism to their advantage.

The numbers of ATP molecules produced by glycolysis and the Krebs cycle are summarized in figure 2.6.

Molecules from other lands can also supply fuel to the Main Powerhouse. Lipids from Lipidland supply much acetyl CoA and glycerol from the Lipidland (triglyceride) storeroom. Tapping the resources of the Lipidland Storage Room occurs all the time, but becomes especially prominent after only a short period of fasting, as the Carbohydrateland storage room has only enough stores (glycogen, which breaks down to glucose) to last about a day. The Amino Acid Midway can supply amino acids which can change into Main Powerhouse molecules, particularly during periods of starvation. However, the Amino Acid Midway really prefers not to be a source of fuel molecules as this means it may have to break down important proteins to get at the amino acids (the Midway does not have its own private amino acid storage room).

The Main Powerhouse is particularly important because it is more than an ATP energy generator that uses carbohydrates, lipids and amino acids as fuel. It can also be the source of a number of molecules that are used in other areas of Biochemistryland. It can generate carbohydrates **(gluconeogenesis)**, lipids **(lipogenesis)** and amino acids. Succinyl CoA is not only used for the Krebs cycle; it is the entry point to Porphy's Hemeland. A number of molecules spun off from the Ferris Wheel (e.g., glutamate and aspartate) are also used to make purines and pyrimidines for the DNA Funhouse. The directions that reactions follow largely depend on the body's needs and on feedback reactions that affect key rate-controlling steps.

The Saloon

If we don't have enough O_2 (as in vigorous exercise) we need some way to replenish the NAD^+ that is used in the Main Powerhouse hallway. As one cannot rely on oxidative phosphorylation to do it, pyruvate transforms into lactate, anaerobically, thereupon replenishing NAD^+. Now a runner who is becoming anaerobic may get a high, but this is not due to pyruvate transforming into alcohol. Only microorganisms can do that. The saloon in Biochemistryland makes alcohol but **only in microorganisms.** If we ingest ethanol we can handle it by transforming it to acetyl CoA. Hence, alcohol can be used as a fuel. Alcohol ingestion can also lead to weight gain, as acetyl CoA connects directly with the lipid-synthesizing roller coaster of Lipidland.

GLYCOLYSIS	**ATP PRODUCTION**
1. Glucose + ATP → Glucose 6-Phosphate + ADP	−1
2. Fructose 6-P + ATP → Fructose 1,6-P_2 + ADP	−1
3. (two) 1,3-P_2-Glycerate + ADP → (two) 3-P-glycerate + (two) ATP	+2
4. (two) P-enol pyruvate + ADP → (two) Pyruvate + (two) ATP	+2
KREBS CYCLE AND OXIDATIVE PHOSPHORYLATION (MITOCHONDRIA)	
5. (two) Pyruvate + (two) NAD^+ → (two) Acetyl CoA + (two) NADH	+6
6. (two) Isocitrate + (two) NAD^+ → (two) 2-ketoglutarate + (two) NADH	+6
7. (two) 2-ketoglutarate + (two) NAD^+ → (two) Succinyl CoA + (two) NADH	+6
8. (two) Succinyl CoA + (two) GDP → (two) Succinate + (two) GTP	+2
9. (two) Succinate + (two) FAD → (two) Fumarate + (two) $FADH_2$	+4
10. (two) Malate + (two) NAD^+ → (two) Oxaloacetate + (two) NADH	+6
GLYCEROL 3-PHOSPHATE SHUTTLE (CYTOSOL ———— MITOCHONDRIA)	
11. IN CYTOSOL: (two) DHAP + NADH → (two) Glycerol 3-P + NAD^+ IN MITOCHONDRIA: (two) Glycerol 3-P + (two) FAD → (two) DHAP + (two) $FADH_2$	+4
TOTAL	+36 ATPs

Fig. 2.6. The numbers of ATP molecules produced in glycolysis and in the Krebs cycle. Thirty six ATPs result from the splitting of one glucose molecule. Note that each 6-carbon glucose splits into two 3-carbon molecules, each of which generates its own ATPs. Each NADH from the Krebs cycle yields 3 ATPs. Each $FADH_2$ yields 2 ATPs. Each NADH from the cytosol yields only 2 ATPs. Each GTP yields one ATP. DHAP, Dihydroxyacetone phosphate.

Where is The Main Powerhouse?

The hallway portion of the Main Powerhouse (glycolysis) through pyruvate lies in the cytoplasm; the remainder (pyruvate through the Krebs cycle) occurs in the mitochondria (fig. 2.3). Physicians generally are not too interested where such cellular partitioning occurs. True, certain diseases may be selective for mitochondria (for instance, certain rare myopathies), but in general, physicians are more interested in which **organ** systems correspond to various portions of the map. Thus, a laboratory test that points to a defect in a particular chemical reaction may provide clues as to which organ is affected. Therefore, regarding **organ** localization, glycolysis is a rather universal reaction, occurring throughout the body. It's reverse, however, gluconeogenesis (all the way back to glucose), occurs predominantly in the liver (to a lesser extent in the kidney). Thus, for lactate to convert back to glucose after exercise, it has to leave the muscle cell, and travel all the way back in the blood stream to the liver, where gluconeogenesis can occur. Once in the liver, the appropriate enzymes are available to help lactate revert to glucose. These enzymes include:

1. **pyruvate carboxylase** (enables pyruvate to become phosphoenolpyruvate by first changing pyruvate to oxaloacetate (D-8). Otherwise the reaction from PEP to pyruvate is essentially irreversible).
2. **fructose diphosphatase**—acts in the changing of fructose 1,6-P2 to fructose 6-P (D-3).
3. **glucose 6-phosphatase,** which is not present in muscle but is present in liver. It catalyzes the change of glucose 6-P to glucose (D-2).

There is a small section of the Main Powerhouse, the toothed section of the key that lies off the main hallway. It contains a side-reaction involving 2,3 P_2-glycerate. This reaction is particularly significant in red blood cells and will be elaborated on further in the Clinical Review #8.

Since red blood cells do not contain mitochondria, they have no Krebs cycle; the main hallway (glycolysis) section does, however, function to produce some of the energy in red blood cells, in addition to the HMP shunt (to be discussed in Carbohydrateland).

Summary of Connections of the Main Powerhouse

1. Carbohydrateland. The Main Powerhouse connects with Carbohydrateland, which provides it with molecules that can be broken down to either be used as fuel (glycolysis) or converted to other molecules. Conversely, the Main Powerhouse, when acting in reverse (gluconeogenesis) can produce carbohydrates.
2. Lipidland. The Main Powerhouse connects with Lipidland, which provides it with acetyl CoA and glycerol, to use as fuel or to convert into other kinds of molecules. Conversely, The Main Powerhouse may provide Lipidland with acetyl CoA and glycerol, to be used in the formation of lipids.
3. The Amino Acid Midway. The Main Powerhouse connects with the Amino Acid Midway, as amino acids, directly or indirectly, can be transformed into Main Powerhouse molecules. Conversely, certain amino acids (the nonessential ones) can be formed directly or indirectly from Main Powerhouse molecules.
4. Porphy's Hemeland. The ferris wheel (Krebs cycle) of the Main Powerhouse connects with Porphy's Hemeland, as the succinyl CoA of the ferris wheel is a precursor of porphyrins and their derivatives, such as heme.
5. The DNA Funhouse. The Main Powerhouse connects indirectly with the DNA Funhouse in that Main Powerhouse ingredients may indirectly become part of purine and pyrimidine nucleotides. Conversely, nucleic acid breakdown may contribute molecules that indirectly transform into Main Powerhouse molecules.

CHAPTER 3. CARBOHYDRATELAND

Carbohydrates are molecules of 3 or more carbons that contain more than one hydroxyl group. In most cases, using the strictest definition of carbohydrates, they also contain an aldehyde group (e.g., glucose) or a keto group (e.g. fructose). Commonly their structure shows one H_2O molecule for each carbon (e.g., glucose is $C_6H_{12}O_6$). Some molecules that are called carbohydrates contain only hydroxyl groups (e.g., mannitol, glycerol), or contain other groups (e.g., glucuronic acid contains a −COOH group).

Carbohydrates include:

1. **Monosaccharides:** carbohydrates that cannot be hydrolyzed (split, with the addition of H_2O) into simpler carbohydrates. Typically, they contain 3,4,5, or 6 carbons (trioses, tetroses, pentoses, hexoses). Glucose (fig. 3.1) as well as fructose and galactose are examples of hexoses.
2. **Disaccharides:** combinations of two monosaccharides. Examples are maltose (glucose + glucose), sucrose (glucose + fructose), and lactose (glucose + galactose). Note: glucose is common to all three of the latter disaccharides. A mnemonic for remembering which names are monosaccharides and which disaccharides: The first letters of the disaccharides (l, lactose; m, maltose; s, sucrose) are later in the alphabet than the first letters of their corresponding monosaccharides (f, fructose; g, galactose; g, glucose).
3. **Oligosaccharides:** contain 3–6 monosaccharides.
4. **Polysaccharides:** contain more than 6 monosaccharides. These include **starches,** which are long chain polymers of alpha-D-glucose in the form of **amylose, amylopectin, and glycogen. Amylose** is an **unbranched** chain of glucose residues connected by alpha 1,4 linkages (fig. 3.1). **Amylopectin,** apart from being a longer word, is more complex than amylose in having additional, alpha-1,6 linkages, which result in branching. **Glycogen**

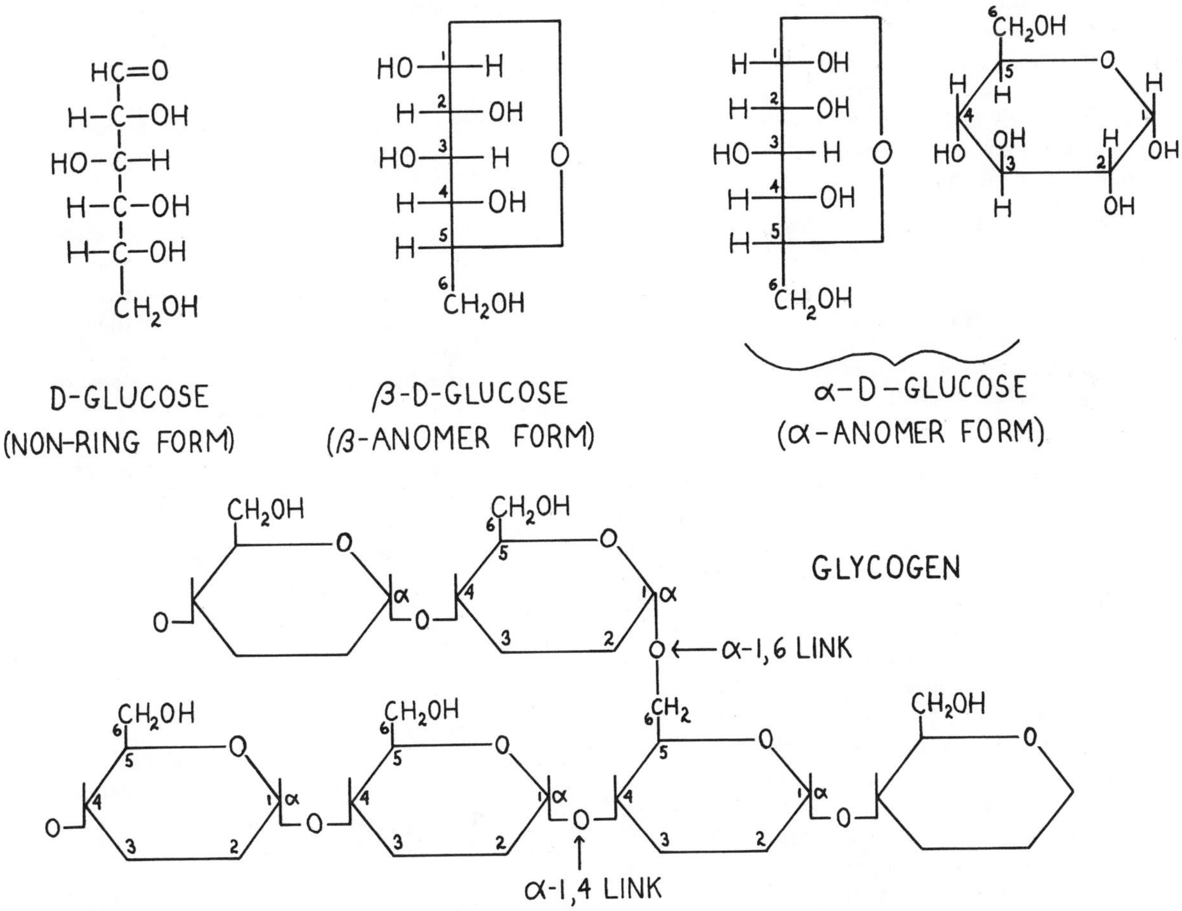

Fig. 3.1. Glucose and its two anomeric forms, which result from the formation of a ring on opening up of the carbonyl (−C=O) group. Note the several ways of drawing molecular structure. Maltose consists of two alpha-glucoses linked by a 1,4 bond. Amylose consists of a chain of such linkages. Glycogen, in addition, contains 1,6 links, which result in a branched structure. See Appendix I, page 73, for further review of isomer terminology.

("animal starch") resembles amylopectin but branches even more. Glycogen is the main storage form of carbohydrates in humans. Branching makes sense. The enzymes that synthesize and break down these long glucose chains like to act at chain end points. Branching allows a multitude of end points for these reactions.

Cellulose has **beta–1,4** linkages, bonds which human enzymes cannot break, but the bacterial enzymes in cows can. Therefore cows eat grass, but we don't.

Carbohydrates have many useful functions. They are used for the storage and generation of energy; they are important structural components, both intra- and extracellar; and they may be transformed into other, totally different kinds of molecules, like amino acids, lipids, and nucleic acids. When carbohydrates attach to proteins and lipids, they form **glycoproteins** and **glycolipids**. The latter complex molecules may be found in Combo Circle and will be seen at show time.

Carbohydrateland connects with the Main Powerhouse. The Main Powerhouse takes Carbohydrateland's molecules and changes them to acetyl CoA, which is used as fuel in the Krebs cycle ferris wheel. Alternatively, the Main Powerhouse can take the same molecules and convert them into lipids and amino acids. Thus, the molecules of both Carbohydrateland and the Main Powerhouse can become fuel or become widely transformed.

Carbohydrateland has some very interesting tourist attractions. Let us explore them further:

The Penthouse Powerhouse

Carbohydrateland has its own powerhouse, called the hexose monophosphate (HMP) shunt, the pentose phosphate pathway, the pentose shunt, or the phosphogluconate oxidative pathway. We'll just call it either the Penthouse Powerhouse, or, briefly, the HMP shunt. This powerhouse does **not** produce ATP energy molecules, as does the Main Powerhouse, but it does produce NADPH, a molecule important for its "reducing" power. Many biosynthetic reactions involve substrate reductions and NADPH is frequently called into play to accomplish this. An important event in the pentose shunt is the release of CO_2 which changes six carbon sugars to a five carbon one—**ribose**. Hurray for that maneuver as it is the ribose that moves along its way to the DNA Funhouse and is an important component of nucleotides (like ATP), DNA and RNA. Ribose is also a part of NADH, FAD and the "CoA" portion of acetyl CoA (see Structural Index). If there is not sufficient demand for the latter items the ribose (as ribose 5-phosphate) can be converted to glyceraldehyde 3-phosphate and used in the Main Powerhouse. In fact, ribose 5-P can do a number of juggling stunts. It can change into 3,4,6 and even 7 carbon sugars. The HMP shunt can even go partly in reverse. For instance, if the body needs ribose 5-P, but doesn't need NADPH, then the shunt can run from glyceraldehyde 3-P back to ribose 5-P, thereby bypassing the NADP-to-NADPH steps from glucose 6-P.

The Carbohydrateland Storage Room

Excess glucose molecules may be stored by linking together to form glycogen, which is stored in the liver, skeletal muscle, and many other tissues. As indicated on the map, the sequence flows from glucose→ glucose 6-P→glucose 1-P→UDP—glucose→amylose (unbranched)→glycogen (branched). Glycogen breakdown requires several kinds of enzymes:

1. **A phosphorylase.** This nibbles off glucose units but cannot break the 1-6 bond, and in fact, stops acting within 4 glucose units of the 1-6 branch point. The glucose that it does chew off is phosphorylated in the process of cleavage to become **glucose 1-P.** The unacted-upon, branched residue **(=limit dextrin)** may be further degraded by:
2. a **debranching enzyme,** enabling the rest of the molecule to be broken down completely, mainly to more glucose 1-P. There are different phosphorylases for muscle and liver, and either may be lacking in certain diseases.
3. In the gastrointestinal tract, salivary and pancreatic **amylase** can break down starch directly to maltose, which in turn is split by maltase (from glands in the small intestinal wall) to form glucose.

The main net product of the phosphorylase reaction (in liver and muscle) is glucose 1-P, whereas the net product of the amylase reaction (in the small intestine) is glucose. There is good reason behind this differential effect. The intestines (amylase) would like to take the low road and form glucose because glucose is readily absorbed from the gut. Glucose 1-P, though, doesn't cross membranes well. Muscle cells like to take the high road and form glucose 1-P as it remains confined within the cell where it can be used, rather than leaking out of the cell, as glucose does. And don't think glucose 1-P could leave the muscle cell so easily by simply changing to glucose 6-P and then glucose; muscle cells (as well as brain cells) lack glucose 6-phosphatase and do not change glucose 6-P to glucose. Muscle cells (as well as brain cells) thus retain their glucose 6-P for use in generating much-needed ATP. Liver cells, on the other hand, do have glucose 6-phosphatase. The liver doesn't care if some of its glucose 6-P changes to glucose. It likes glucose to leave its cells. The liver, in fact, is a major supplier of glucose to the rest of the body, itself preferring other molecules as fuel, like lactate and fatty acids.

By using a phosphorylase or amylase, glycogen may be broken down to glucose-1-P or to glucose. Glucose 1-P is not formed by glycogen changing back to UDP-glucose to glucose 1-P. An important clue that would predict this unlikelihood is the production of the double inorganic phosphate, pyrophosphate (PPi, see Structural Index) in the reaction of glucose→UDP-glucose. PPi rapidly combines with water to form two inorganic phosphate (Pi) ions, a virtually irreversible reaction. Hydrolysis of PPi insures that many reactions in the body procede in a one-way direction (e.g. PRPP→5-P-ribosyl amine, map section B-4).

Insulin is an anabolic hormone that "signals the fed state". It reacts to feeding by clearing the blood of glucose, storing the fuel, and promoting glycogen, fatty acid, and protein synthesis. It stimulates **glycolysis** (breakdown of glucose) and inhibits **gluconeogenesis** (formation of glucose). It also facilitates entry of glucose into muscle and fat cells.

Epinephrine and glucagon are hormones that promote glycogen breakdown, in a sense acting opposite to insulin. Epinephrine is more effective in muscle whereas glucagon is more effective in the liver. Glucagon and epinephrine restore blood glucose levels by enhancing glycogen breakdown, decreasing glycogen synthesis, decreasing glycolysis and fatty acid synthesis and stimulating gluconeogenesis. Epinephrine and glucagon do not enter the cell but act at the cell membrane level by stimulating the enzyme **adenylate cyclase.** This causes a cascading reaction, in which cyclic AMP acts within the cell as a second messenger that leads to the activation of phosphorylase (and inhibition of glycogen synthase). This stimulates glycogen breakdown (and inhibits glycogen synthesis) (fig. 3.2 and map section D-1).

Insulin, like epinephrine and glucagon, appears to act on a receptor on the cell membrane, but does not appear to use cyclic AMP as a second messenger.

Fructose and galactose absorbed by the intestines also enter Carbohydrateland. Fructose may form fructose 1-P, which splits into two halves as does glucose, to form two 3-carbon molecules (D-4). Galactose connects with the Main Powerhouse through its conversion to UDP-glucose and thereby can be used as a fuel source.

The Ice Cream Parlor

Lactose (a prime ingredient in milk and ice cream) is a disaccharide that may be found in the Ice Cream Parlor, in addition to the monosaccharide, galactose. Lactose and galactose may be produced from glucose via UDP-glucose. Galactose is a particularly important component of complex glycoproteins and glycolipids. Galactose can not only be synthesized when absent from the diet, but can also convert back to glucose should the need arise for it to be used as a fuel.

Conjugation Area

Now don't let your imagination run away with you. The Conjugation Area is not an illicit area of Carbohydrateland. If you want that sort of thing you had best visit the sexual zone of Sterol Forest in Lipidland. The Conjugation Area produces glucuronate, one of the important molecules that can join (conjugate) with drugs, thereby inactivating them and facilitating their excretion. Glucuronate plays a major role in the conjugation and excretion of bilirubin. This will be discussed further when we visit Porphy's Hemeland. Note the pathway that extends off the western side of the map from glucuronate and enters Porphy's Hemeland on the eastern side of the map.

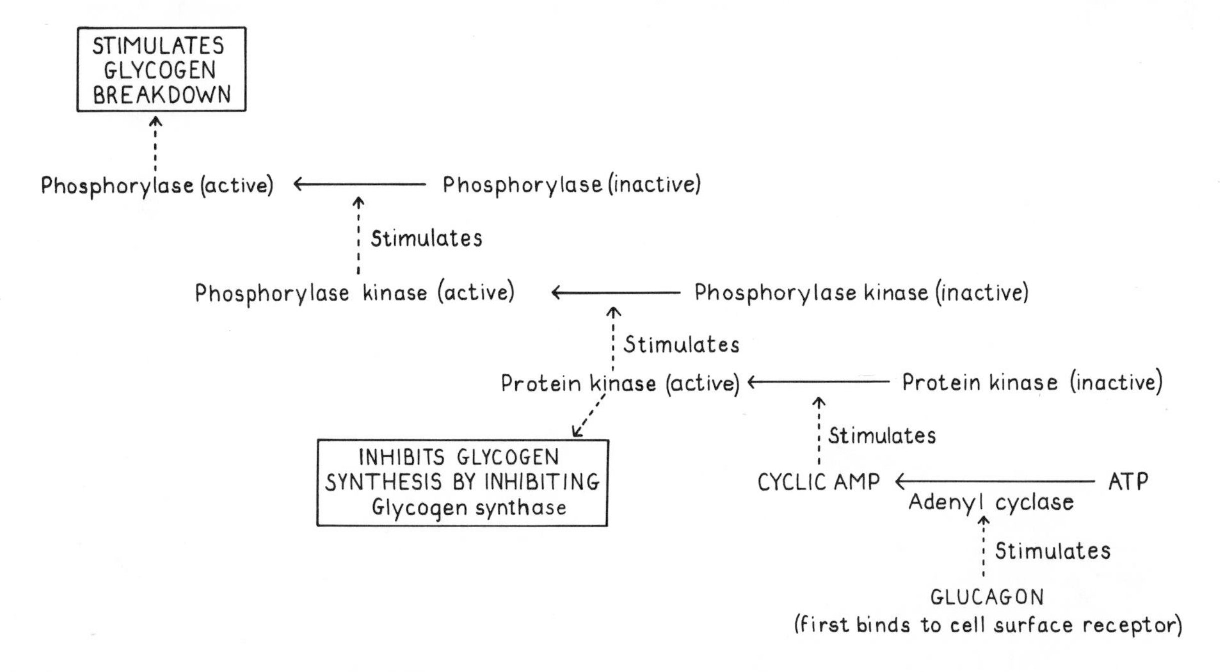

Fig. 3.2. The cyclic AMP cascade in phosphorylase activation, in the liver.

Where is Carbohydrateland?

The Penthouse Powerhouse (HMP shunt) is found throughout the body. The shunt is a particularly important source of reducing power in red blood cells as these cells lack mitochondria, the normal site of the Krebs cycle. Red blood cells use the reducing power of NADPH to keep the iron atoms of hemoglobin in the reduced (Fe^{++}) state. NADPH also protects red cells against potentially harmful oxidants. The HMP shunt is very important in fat and liver cells which use NADPH quite a lot for fatty acid synthesis and (in liver) for cholesterol synthesis. The adrenal cortex also uses significant amounts of NADPH for steroid hormone synthesis.

Carbohydrateland lies partly in the gut, too, Salivary and pancreatic amylase digest starches in the digestive tract. The products are glucose and disaccharides that are absorbed and, after further hydrolysis in the intestinal cells, directly go to the liver via the blood stream. The small intestinal wall also produces sucrase (which changes sucrose to glucose and fructose), lactase (which changes lactose to glucose and galactose), and maltase (which splits maltose into two glucose molecules). Fructose and galactose enter the small intestinal epithelial cells where they may be changed to glucose or enter the bloodstream along with glucose.

Liver and muscle cells contain phosphorylases which break down starches somewhat differently than in the gut (by phosphorolysis rather than hydrolysis), forming glucose 1-P in preference to glucose, as discussed above.

Lactose, seen in the Ice Cream Parlor, is synthesized preferentially in the mammary glands. Lactase, which facilitates the breakdown of lactose to glucose and galactose, is found in microvilli of the small intestine. Glucuronate synthesis occurs largely in the liver where it is used to conjugate bilirubin and to detoxify various chemicals in preparation for excretion.

Summary of Key Connections of Carbohydrateland

1. The Main Powerhouse. Carbohydrates are split into 3-carbon groups that enter the Main Powerhouse to be used as fuel (through acetyl CoA) or converted into other molecules such as lipids (through glycerol and acetyl CoA), and amino acids. Carbohydrateland connects with the Main Powerhouse through glyceraldehyde 3-P and dihydroxyacetone. Fructose connects through glyceraldehyde 3-P, fructose 1-P, and dihydroxyacetone, which provide a link to Lipidland as well, via glycerol. Galactose connects through UDP-glucose. Reversibility of the above reactions is important when the need arises for particular molecules.
2. The DNA Funhouse. Ribose from the HMP shunt provides the sugar group that is part of purine and pyrimidine nucleotides.
3. The Combo Theatre—specifically the Glycoprotein and Glycolipid sections. Fructose 6-P combines with glutamine to form the amino sugar, glucosamine 6-P (E-2). The latter then transforms into other amino sugars which, together with simpler carbohydrates, make an important contribution to glycoprotein and glycolipid structure.

If you are in the mood for hotdogs and hamburgers, let us now visit Lipidland.

CHAPTER 4. LIPIDLAND

Lipids are a very diverse group of biochemicals that, by definition, are poorly soluble in water and are soluble in organic solvents such as ether. An important structural feature of lipids that allows these solubility characteristics is a relatively high ratio within the molecule of hydrocarbon atoms to more polar kinds of atoms. This definition allows the inclusion of tons of molecules into the lipid family, even gasoline. We will focus on the ones normally found in the body.

Lipids have a number of important functions:

1.**Fuel.** Fatty acids (R−COOH), on breakdown, form acetyl CoA, NADH, and $FADH_2$, important sources of energy. When attached to glycerol, fatty acids may be stored as **triglycerides (fat).**
2. **Cell membranes.** Lipids, particularly the phospholipids, glycolipids, and cholesterol are important cell membrane components throughout the body. Sphingomyelin, cerebrosides, and gangliosides are particularly important cell membrane constituents in the nervous system.
3. **Steroid hormones** are lipids that have the characteristic sterol ring, part of which looks like a house (fig. 4.1). The side chains attached to the C17 carbon are the main determinants of steroid activity. Pregnenolone, the first steroid hormone derivative of cholesterol, gives rise to a number of different steroids:
A. **Sex hormones** (e.g., estrogens, progesterone, testosterone)

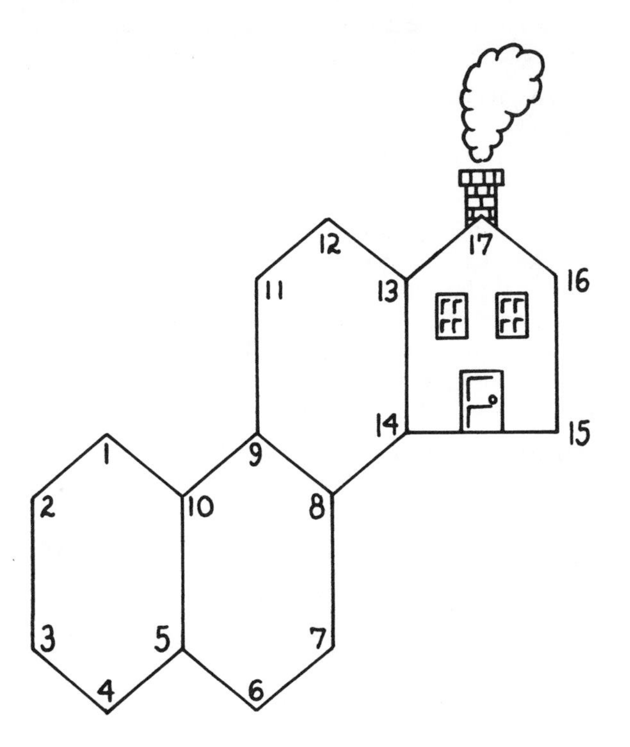

Fig. 4.1. The basic sterol structure.

B. **Glucocorticoids** (e.g., cortisol). These cause a rise in serum glucose level.
C. **Mineralocorticoids** (e.g., aldosterone). These retain sodium in the body.

Sex hormones, glucocorticoids and mineralocorticoids are discussed in greater detail in the section on hormones, in the Infirmary.

4. **Bile acids** are lipids. They are steroids that have a free—COOH group. Their part polar and part non-polar characteristics facilitate their functioning as detergents which bind to both lipids and to the surrounding polar medium, thereby emulsifying fats and increasing their surface area for further breakdown in the gut. **Glycocholate** is the main bile salt.
5. **Prostaglandins** are froggy-looking lipids with diverse hormone-like functions.
6. The **fat-soluble vitamins** (A,D,E,K,) are lipids.
7. Lipids can combine with carbohydrates or proteins to form **glycolipids** and **lipoproteins.** These will be examined in greater detail in Combo Circle.

Fatty acids are especially important and ubiquitous lipids. Apart from their use as fuel, they may be used in the synthesis of many other kinds of molecules. A fatty acid is a hydrocarbon chain with a terminal −COOH group. Usually, fatty acids contain between 14 and 25 carbons, 16 and 18 being most common. **Unsaturated** fatty acids are those that contain double bonds between some carbon atoms. **Saturated** fatty acids contain only single bonds. Palmitic acid (16) and stearic acid (C18) are particularly common saturated fatty acids in humans. Linoleic acid (C18), linolenic acid (C18) and arachidonic acid (C20) are common polyunsaturated fatty acids ("poly" meaning more than one double bond). The number of carbons typically is even, reflecting the fact that fatty acid chains are built up (and broken down) in units of 2 carbon atoms at a time.

Let's begin by visiting the Up and Down Roller coasters in the CENTRAL SECTOR (middle hot dog) of Lipidland.

The Up and Down Roller Coasters

It is illogical to think of a roller coaster that goes only up or down. Nor do the terms "up" and "down" refer to one's experiencing an emotional high on one and a low on the other. The terms refer to the processes of lipid synthesis (up) and degradation, or oxidation (down).

The peculiar spiral nature of the roller coasters is due to the repetitive cycles involved in lipid synthesis and degradation. In other words, lipid synthesis and degradation occur by adding or removing 2 carbons with each cycle. Imagine a roller coaster train in which the train gains 2 seats every time it completes a cycle. The 2 seats are carbons, gained as an acetyl group from acetyl CoA with each round of synthesis, thereby extending the fatty acid chain. Similarly, lipid oxidation consists of the loss of 2 carbons with each go-around. The seats of the up and down roller coasters can be distinguished from one another in that the seats of the synthesis roller coaster are affixed to acyl (fatty acid) carrier protein (ACP). The oxidation roller coaster seats attach to CoA.

The breakdown of fatty acids is an important source of energy. For one thing, NADH and $FADH_2$, which are energy molecules, are generated. Moreover, acetyl CoA is formed, an important fuel molecule for the Krebs cycle ferris wheel. Each round of 2 carbon losses produces its own NADH, $FADH_2$ and acetyl CoA. Considerable energy can be generated with successive rounds of 2-carbon losses.

Lipids are more energy efficient molecules than carbohydrates, supplying 9 kcal/g, as opposed to 4 kcal/g for carbohydrates and proteins. Part of the reason for the high energy potential is that considerable energy is released in the multiple cycles of fatty acid oxidation; triglycerides are also relatively non-polar when compared with proteins and carbohydrates and therefore bind less water. This dehydrated state allows for relatively compact packaging of triglycerides.

When blood sugar is low, epinephrine, norepinephrine and glucagon stimulate a lipase in the fat cells to break down triglycerides, the glycerol part of which can be used to produce more glucose. Cyclic AMP, discussed previously in Carbohydrateland, acts as a second, intracellular messenger in carrying the hormonal message that eventually activates this intracellular lipase, much as cyclic AMP assists in carrying the message for phosphorylase activation in glycogenolysis.

Insulin inhibits lipid breakdown by a number of mechanisms and increases synthesis of glycogen, fatty acids, triglycerides, and proteins. Insulin reduces the level of cyclic AMP. It also promotes the transport of glucose into cells, especially fat cells, where glucose can be converted and stored as triglyceride.

Not shown on the map are certain modifications for dealing with unsaturated and odd chain fatty acids. We cannot synthesize all fatty acids. Linoleate (C18) and linolenate (C18) are **essential** fatty acids that are required in the diet, as we cannot synthesize their double bonds.

Ketone Playhouse

It's now time for a performance of that energetic musical group, **The Ketones!** So if you're hongry and tired, get yourselves down to the Ketone Concert Stage for a little refreshment and entertainment.

Ketones are produced in the course of breakdown of fatty acids. At some point the breakdown reaches the point where the fatty acid is degraded to the 4-carbon acetoacetyl CoA. Acetoacetyl CoA can either:

1. break down further to acetyl CoA,
2. change to ketones (**acetoacetate (C4), hydroxybutyrate (C4),** and **acetone** (C3), fig. 4.2),
3. be used for synthesis of cholesterol and its many derivatives in the eastern sector of Lipidland.

Ketones commonly are elevated in the blood in states of starvation, as the body calls upon its fatty acids (stored as triglycerides) to break down and provide fuel. Ketones may also be elevated in diabetes mellitus, where glucose does not enter the cell and cannot be efficiently utilized. Triglycerides then break down to provide the fatty acids and acetyl CoA useful as fuel, sometimes with formation of ketones as well. In severe diabetic ketosis, one may actually detect the smell of acetone coming from the patient.

Ketones themselves may be used as fuel. The brain, which normally prefers glucose, can use ketones in states of starvation. Cardiac muscle commonly uses ketones as fuel.

Now let's move to the WESTERN SECTOR of Lipidland and take a look at the Lipid Storage Room:

The Lipid Storage Room

Fatty acids are stored in this room (i.e., in fat cells) on glycerol coat racks (fig. 4.3) The three −OH groups of glycerol are the hooks of the 3-carbon coat rack. The fatty acids are the clothes that are placed on the coat racks. The products are called **glycerides,** namely:

1. monoglycerides (contain one fatty acid group on the coat rack)
2. diglycerides (two fatty acid groups)
3. triglycerides (three fatty acid groups)

The hand that holds the fatty acid up to the coat rack is CoA. Acyl CoA means a fatty acid attached to CoA. Acyl CoAs can combine (indirectly) with glycerol to form triglycerides. Triglycerides are the primary storage form of lipids. They can revert to fatty acids and glycerol, which in turn can be used as fuel in the Main Powerhouse, or they can change into other kinds of molecules.

Triglycerides thus can be formed from glycerol and acetyl CoA, which are both spun off from the Main Pow-

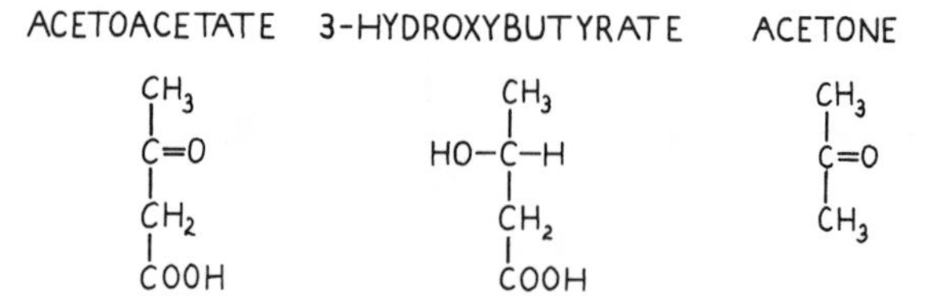

Fig. 4.2. The ketones.

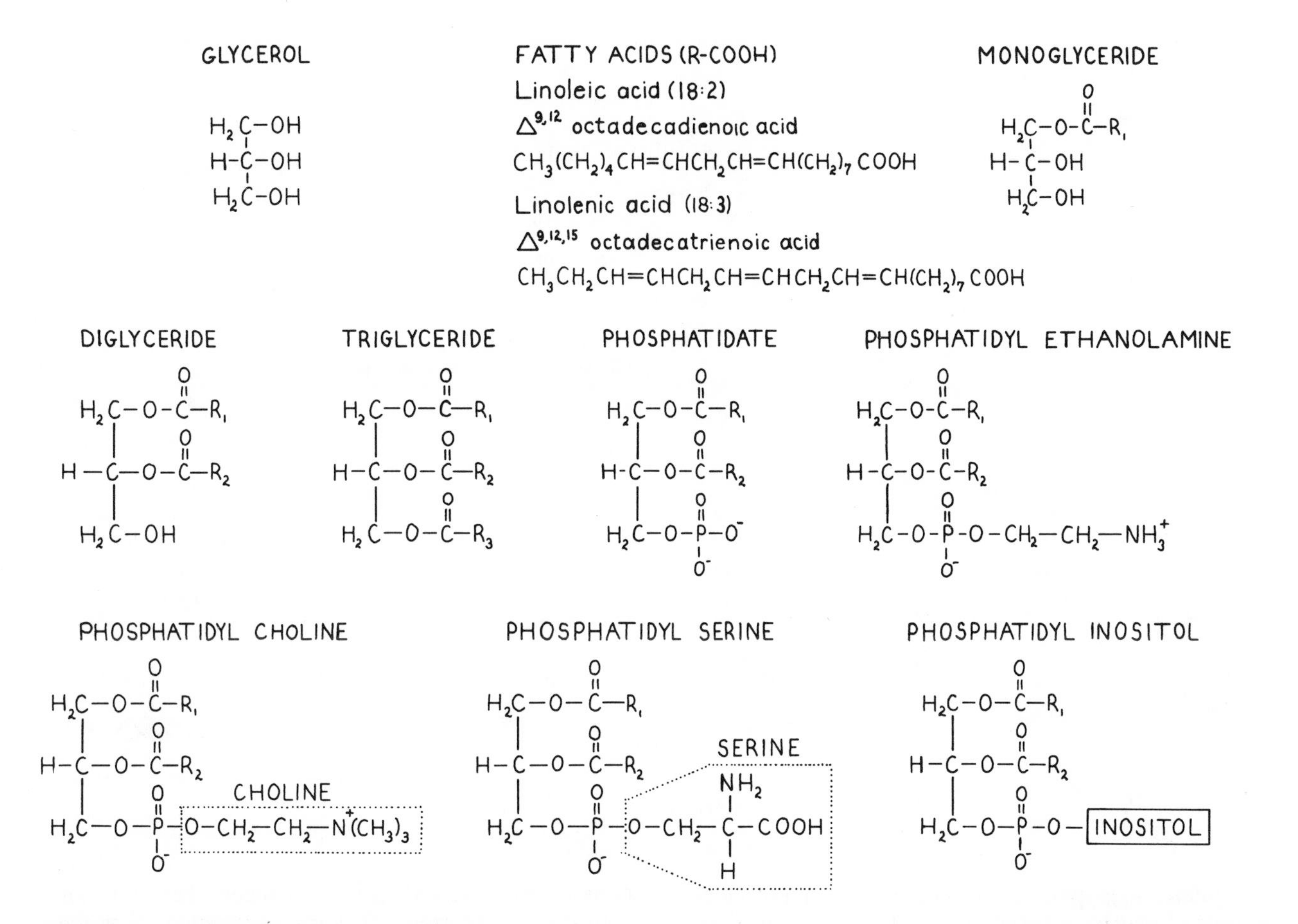

Fig. 4.3. The glycerol molecule (coatrack) and the fatty acid molecule (clothes). Fatty acids attach to the glycerol molecule to form glycerides. When other groups attach to the lower carbon, via a phosphate linkage, one gets phosphoglycerides (the inhabitants of Phosphatidylywink Village).

erhouse. No wonder we get fat on eating too many candy bars, or for that matter, any food that can enter the Main Powerhouse and supply calories. The Main Powerhouse connects with Lipidland. Look where the excess sugar goes—into glycerol, fatty acids (via acetyl CoA) and then triglycerides, which are stored in fat cells.

Phosphatidylywink Village (Phosphoglycerides)

Other kinds of groups can also fit on the glycerol coat rack, such as phosphate, serine, ethanolamine, choline and inositol (fig. 4.3). In general, these groups attach to the free (third) -OH group of diglycerides via a phosphate linkage (hence, the term phosphoglycerides). Phosphoglycerides are important membrane components. They are the residents of Phosphatidylywink Village. Phosphatidyl choline (lecithin) is a starting point for the synthesis of many of the prostaglandin-like molecules.

The Frog Pond (Prostaglandins)

Prostaglandins are froggy-looking fatty acids that contain 20 carbons arranged as a 5-carbon ring with 2 legs, one of which contains a −COOH group. (fig. 4.4). They and molecules of related structure—the **thromboxanes** and **leukotrienes**—are derived from polyunsaturated fatty acids via arachidonic acid (fig. 4.4). Prostacyclins have an extra ring. Thromboxanes have an oxygen added to the 5-carbon prostaglandin ring. Leukotrienes do not have an enclosed ring. Prostaglandins are synthesized throughout the body and have numerous hormone-like effects, depending on the specific prostaglandin. Some prostaglandins may have effects opposite to one another. A few of the functions affected include:

1. **Smooth muscle contraction.** Affected are blood pressure, blood flow, the degree of bronchial constriction, and uterine contraction.

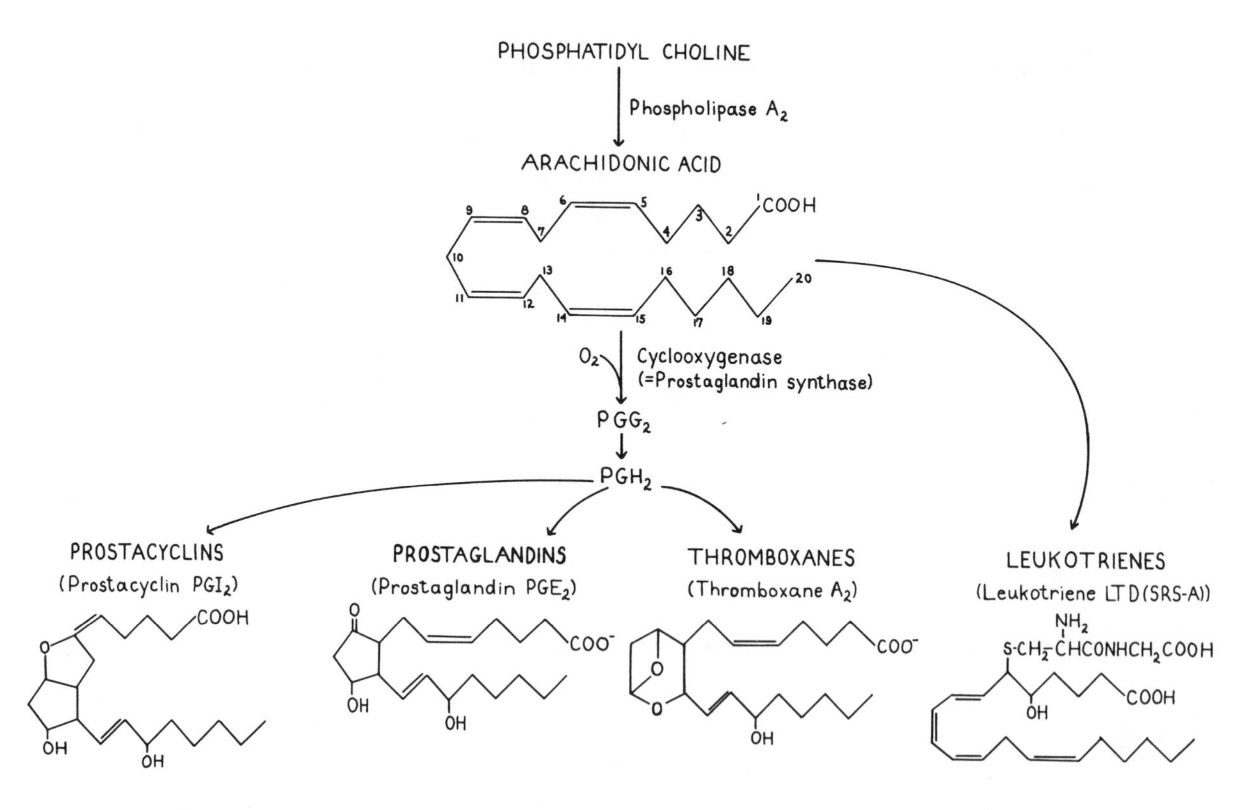

Fig. 4.4. Prostaglandins, thromboxanes, and leukotrienes.

2. **Platelet aggregation.** Thromboxane is a potent **enhancer** of platelet aggregation. Prostacyclin is an **inhibitor** of platelet aggregation as well as a vasodilator.
3. **The inflammatory response.** Certain prostaglandins promote the inflammatory response. The leukotrienes act as chemotactic agents, attracting leukocytes to the site of inflammation.
4. Certain prostaglandins appear to increase pain and fever, perhaps acting in part on certain regions of the hypothalamus.

Sphingo's Curio Shoppe

Sphingo's Curio Shoppe is a delightful little place that is well-stocked with elaborate and unusual items. Sphingolipids look somewhat like triglycerides except for several things. The coat rack is made from serine (a three-carbon amino acid) rather than glycerol (fig. 4.5). This means that the coat rack has an -NH2 group (rather than an -OH group) in its middle carbon, and, initially, a -COOH group (rather than an -OH group) on its first carbon. The -COOH group is linked with a direct carbon-carbon bond to palmitoyl CoA to form **sphingosine** (fig. 4.5). The long saturated fatty acid (palmitic acid) in the latter linkage provides the justification for placing Sphingo's Curio Shop in Lipidland. Then, there is even greater justification when a second fatty acid joins with the −NH2 group, to form a **ceramide.** Now there are two fatty acids joined to the coat rack. This leaves the last carbon on the coatrack with an −OH group that can combine with a whole host of things, forming a whole subclassification of sphingolipids. For instance:

1. **Sphingomyelin** has a phosphoryl choline attached to the third carbon −OH. It is the only phosphosphingolipid.
2. Sugars may also attach to the third carbon −OH group. When that happens, we are talking about **glycolipids,** a topic to be discussed further when in the glycolipid section of Combo Circle. Glycolipids are important components of cell membranes.

The Eastern Sector Of Lipidland

The broad definition of lipids allows the inclusion of cholesterol, a molecule which contains many hydrocarbon groups and is not too soluble in water. Cholesterol contains a skeleton **sterol** ring (fig. 4.1), and is the precursor of other sterol-containing molecules, including the steroid hormones and bile acids (H-11). Cholesterol is ingested in the diet but may also be synthesized. All the carbons of cholesterol come from acetyl CoA. Acetyl CoA units

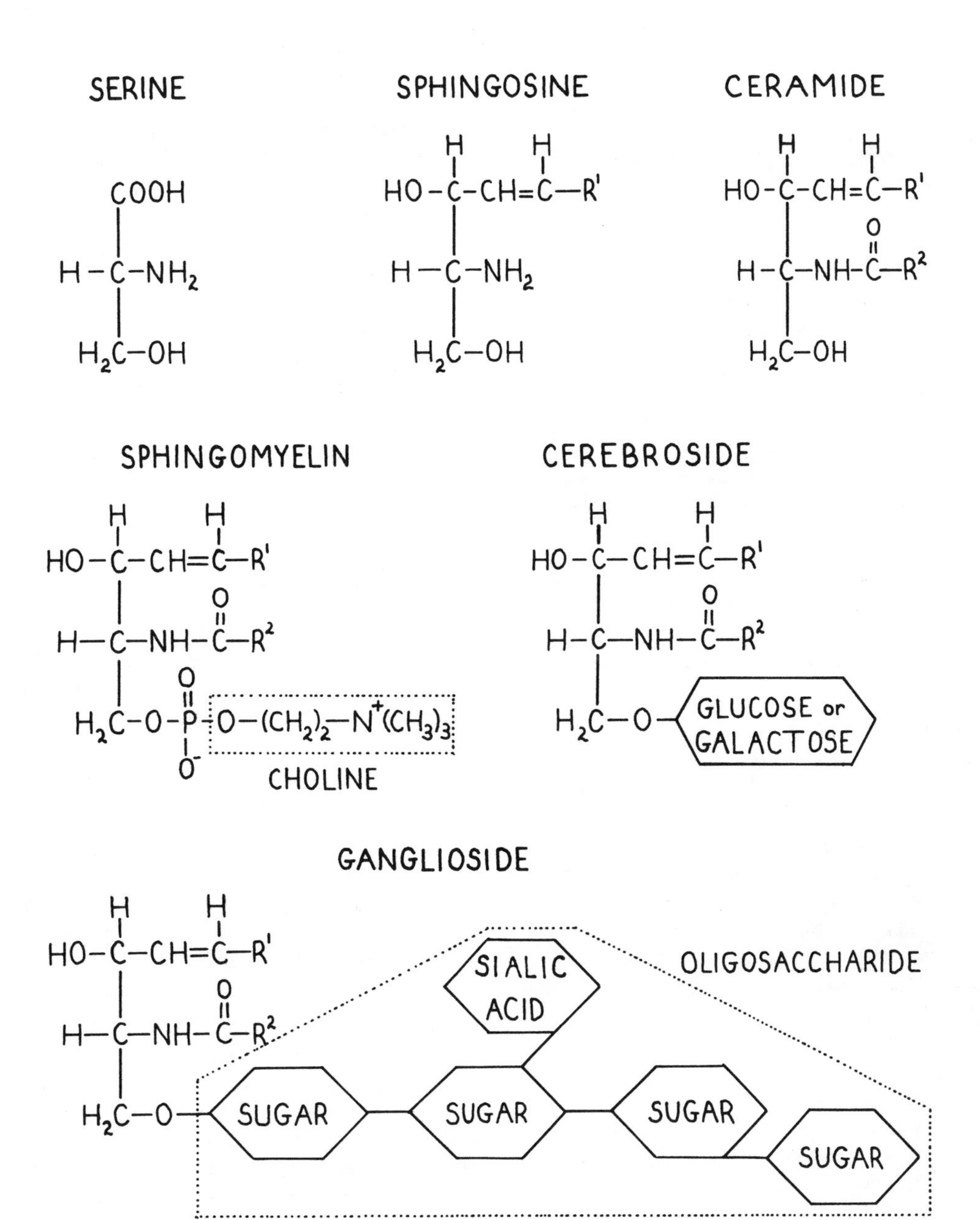

Fig. 4.5. Sphingo's Curio Shop. The coat rack in this case is the 3-carbon serine molecule, rather than glycerol. Fatty acids may attach just to the top carbon **(sphingosine),** or to both the top and middle carbons **(ceramide).** When, in addition, other groups attach to the lower carbon, one may get **sphingomyelin** (phosphocholine attaches), **cerebrosides** (single glucose or galactose attach), or **gangliosides** (oligosaccharide with sialic acid attach).

combine to form the 6-carbon molecule, 3-hydroxy-3-methyl-glutaryl CoA, and then a 5-carbon building block (an **isoprene**—fig. 4.6) which links up in multiple units with other 5-carbon isoprene units. The molecules that are multiples of this 5-carbon basic unit are called **isoprenoids** and may be found in Channel No. 5. The C5 molecule becomes C10,C15,C30, and finally, with some last minute modifications, cholesterol, which has 27 carbons.

The isoprenoid molecules in Channel No. 5 are a sexy grouping, as they give rise to:

1. The terpenes (in plants), which are the active factors in many perfumes (especially the C10 and C15 molecules). Could these be the ingredients in the real Chanel #5?
2. Vitamin E, which may play a role in sexual functioning
3. Chlorophyll (breath mints)
4. Rubbers

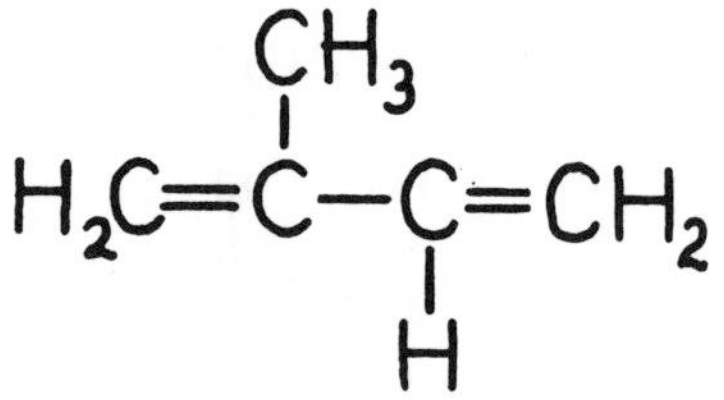

$$H_2C=\underset{}{\overset{CH_3}{\overset{|}{C}}}-\underset{\underset{H}{|}}{C}=CH_2$$

Fig. 4.6. The isoprene building block that constructs the molecules of Channel #5 and Sterol Forest.

In addition to vitamin E, the isoprenoids and their derivatives up to cholesterol give rise to all the fat-soluble vitamins (A,D,E,K) and to coenzyme Q (used in electron transport in the Ferris Wheel Generator). Dolichol is an isoprene polymer that participates in the transfer of oligosaccharides during glycoprotein synthesis.

Cholesterol is not just a greasy molecule that causes heart attacks. It is an important component of biological membranes. Like the isoprenoids, it is also a sexy molecules, giving rise to the sex hormones, among other things.

Interestingly, the need for O_2 in various points in Biochemistryland arises, generally, relatively far out on the individual pathways. Anaerobic glycolysis, for instance, precedes the aerobic aspect of glucose metabolism. Oxygens are added near cholesterol and beyond in Lipidland. This has supported the hypothesis that in the early stages of evolution O_2 was not an important component of the atmosphere but later was formed and became incorporated in reactions that were added on to the preexisting anaerobic ones. The cytochrome system, as described previously (fig. 2.5), serves as an electron transfer chain that results in O_2 incorporation during oxidative phosphorylation. A different cytochrome, **cytochrome P450,** facilitates the incorporation of O_2 in a different kind of reaction—**hydroxylation,** which occurs at a number of points in the synthesis of cholesterol and its derivatives, where O_2 is incorporated in the reaction. Hydroxylation, through P450, has clinical relevance. Hydroxylation is used to detoxify certain drugs (like barbiturates). On the other hand, it may change certain potentially carcinogenic agents into actively carcinogenic ones.

Once beyond cholesterol, one can proceed, if one wishes, along the road to the sexual area of Sterol Forest, which lies far to the East, but one must beware of falling into Bile Bog. A trip through Sterol Forest begins with cholesterol, from which stem all of the steroid hormones, including the mineralocorticoids, glucocorticoids and sex hormones. Cholesterol also gives rise to the bile salts in Bile Bog. The safest way to avoid Bile Bog is to step lively to pregnenolone, the first hormone produced by cholesterol and an important crossway to all the steroid hormones (fig. 4.7).

The various side chains attached to the C17 carbon of the sterol ring are the main determinates of hormonal action, although variations on other parts of the ring also have modifying effects on hormonal activity. In general, there are fewer carbons on the steroid molecule as one progresses from cholesterol. For instance, cholesterol is C27; pregnenolone, progesterone, glucocorticoids, and mineralocorticoids are C21, androgens are C19, and estrogens are C18. Estrogens differ in part from androgens as they lack a certain C19 appendage that is present on the androgen molecule. Estrogens, while not smelling as nice as the perfumes in Channel No. 5, are the only steroid hormones to contain an aromatic ring.

The steroid hormones will be discussed further in the Infirmary chapter and in the Clinical Review.

Where Is Lipidland?

Fatty acid synthesis occurs in the cytoplasm, whereas fatty acid oxidation occurs in mitochondria. As mentioned previously, physicians are not so much interested in the intracellular localization of reactions as they are in the distribution of the reactions in the various organ systems. Lipids are such important components of cell membranes that the processes of lipid biosynthesis and degradation are near universal. Lipid **storage** as triglycerides, however, is mainly a function of fat cells.

Some of the elaborately produced sphingolipids concentrate mainly in the nervous system (e.g. sphingomyelin in the myelin sheath).

The liver is the most important area for the production of cholesterol and bile acids. The ovaries are the most important sources of estrogen and progesterone (progesterone is produced in the corpus luteum of the ovary). The testes are the major site of androgen production, testosterone especially. The adrenal cortex produces the mineralocorticoids and glucocorticoids, and the androgens dehydroepiandrosterone and androstenedione.

As cholesterol is used in cell membranes and in the formation of the steroid hormones, and the liver is the most important source of cholesterol synthesis, the liver exports much cholesterol in lipoprotein trams that carry cholesterol (as well as triglycerides) to other areas of the body, where it is used in further syntheses. The liver also exports cholesterol and bile salts into the bile.

Dietary triglycerides and esterified cholesterol and phospholipids are broken down in the intestine through, respectively, pancreatic lipase, cholesterol ester hydrolase, and phospholipase A2, to form free fatty acids, unesterified cholesterol and lysophospholipids (phospholipids with only one fatty acid group), which are absorbed.

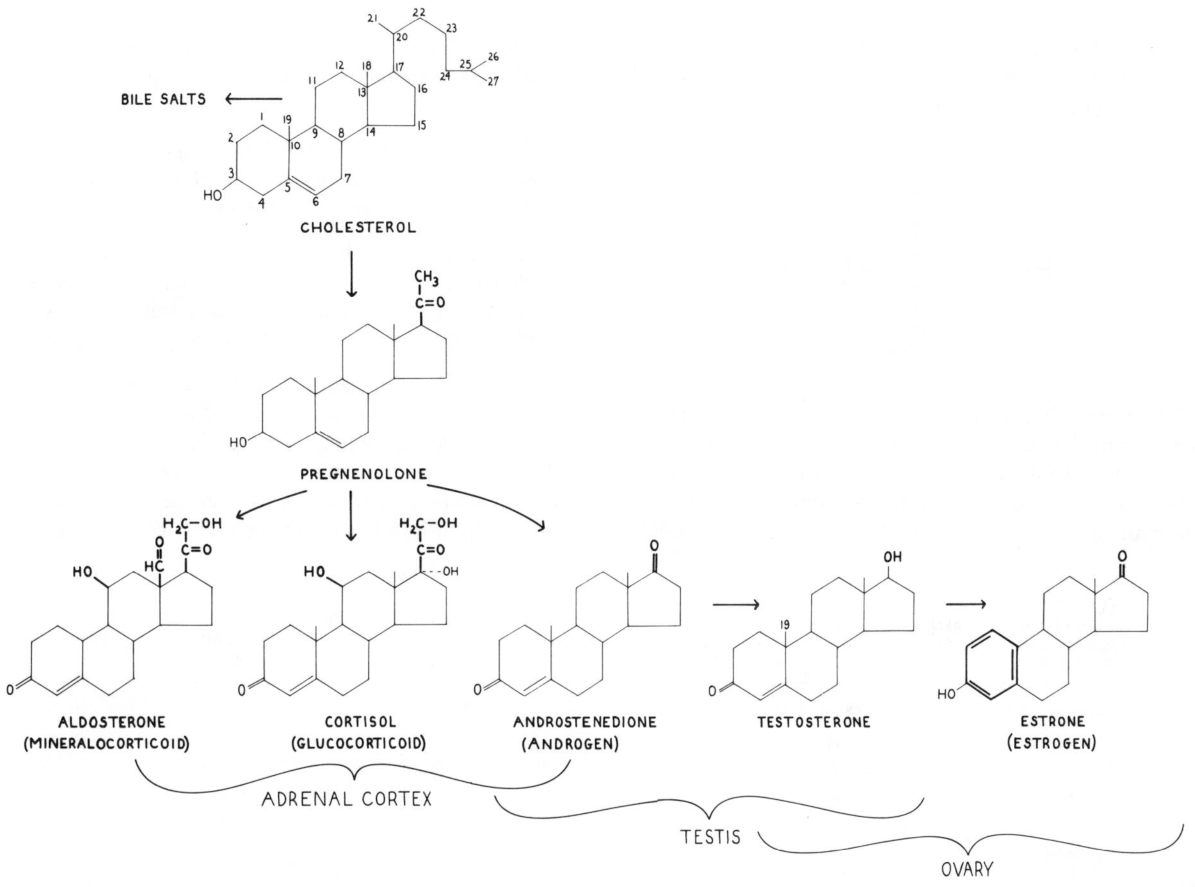
BILE SALTS
HO
CHOLESTEROL
CH3
C=O
HO
PREGNENOLONE
H2C-OH
C=O
HC
HO
O
H2C-OH
C=O
OH
HO
O
O
O
OH
19
O
O
O
HO
ALDOSTERONE
(MINERALOCORTICOID)
CORTISOL
(GLUCOCORTICOID)
ANDROSTENEDIONE
(ANDROGEN)
TESTOSTERONE
ESTRONE
(ESTROGEN)
ADRENAL CORTEX
TESTIS
OVARY

Fig. 4.7. Steroid hormone derivatives of cholesterol.

The brain does not use fatty acids as a fuel, as fatty acids are transported in the blood bound to albumin and do not cross the blood brain barrier. The brain prefers glucose as fuel, but does not manufacture it. Instead, it acquires it from the diet or from the liver. In starvation, the brain may use ketones as fuel. The liver largely obtains its energy from lactate (after a meal) or from fatty acids (after fasting).

Summary Of Key Connections Of Lipidland

1. The Main Powerhouse. Glycerol from the Main Powerhouse forms the coat rack backbone of the triglycerides. Acetyl CoA from the Main Powerhouse forms fatty acids in the up roller coaster. Conversely, triglycerides can break down to glycerol and acetyl CoA, important fuel sources for the Main Powerhouse. Fatty acids cannot form glucose, as the acetyl CoA that fatty acids make when they break down cannot procede to glucose through gluconeogenesis. However, even though fatty acids cannot form glucose, the energy they release from breakdown can be used to help drive gluconeogenesis. Also, the glycerol derived from triglyceride breakdown can form glucose even though the fatty acids cannot.
2. The Amino Acid Midway. Certain amino acids (called **ketogenic**) can be converted to ketones, which in turn can be used for energy or further syntheses.
3. Carbohydrateland—through the formation of glycolipids, but see Combo Circle for this, as well as for lipoproteins.

CHAPTER 5. THE AMINO ACID MIDWAY

The alpha-amino acids contain an amine ($-NH_2$) group, an acidic carboxyl ($-COOH$) group and a variable $-R$ (hydrocarbon-containing group attached to a central carbon:

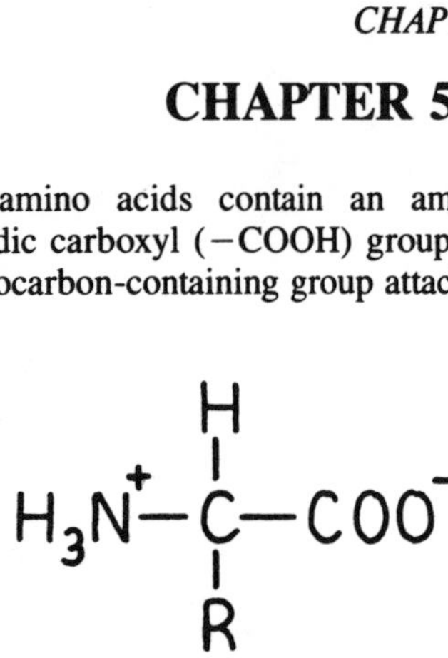

$$H_3N^+-\underset{R}{\overset{H}{C}}-COO^-$$

The term "alpha-amino" means that the amino group is attached to the first (alpha) carbon of the chain (the carbon of the $-COOH$ group does not count in the numbering). An $-R$ group may simply be composed of carbon and hydrogen (a hydrocarbon chain), but may also contain added $-COOH$, $-NH_2$, $-OH$, sulfur groups, or rings, such as a 6-carbon aromatic ring or an indole or imino ring (fig. 5.1).

The amino acids do not have a special land. They are scattered throughout the Midway because of their extensive interconnections with the other lands.

Amino acids are best known as the building blocks of proteins, and there are some 20 kinds of amino acids in proteins (fig. 5.1). There are many other kinds of amino acids, though, that are not found in proteins (fig. 5.2).

The "essential" amino acids are those that the body does not synthesize in significant quantities; they are required in the diet. Plants produce these amino acids, partly through nitrogen-fixing bacteria in the soil. The essential amino acids generally require complex biosynthetic pathways which humans don't have but which many bacteria

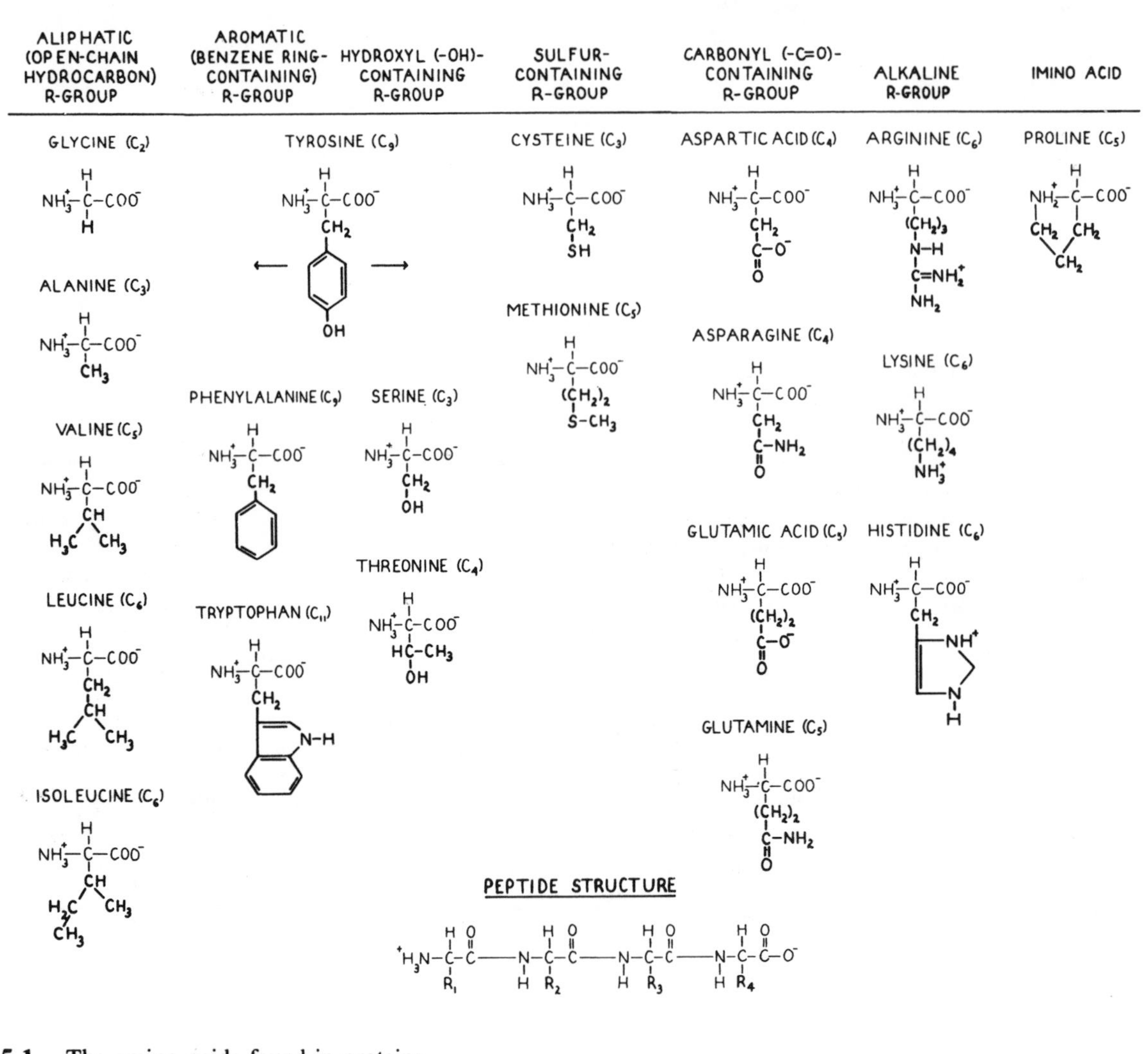

ALIPHATIC (OPEN-CHAIN HYDROCARBON) R-GROUP: GLYCINE (C_2), ALANINE (C_3), VALINE (C_5), LEUCINE (C_6), ISOLEUCINE (C_6)

AROMATIC (BENZENE RING-CONTAINING) R-GROUP: TYROSINE (C_9), PHENYLALANINE (C_9), TRYPTOPHAN (C_{11})

HYDROXYL (-OH)-CONTAINING R-GROUP: SERINE (C_3), THREONINE (C_4)

SULFUR-CONTAINING R-GROUP: CYSTEINE (C_3), METHIONINE (C_5)

CARBONYL (-C=O)-CONTAINING R-GROUP: ASPARTIC ACID (C_4), ASPARAGINE (C_4), GLUTAMIC ACID (C_5), GLUTAMINE (C_5)

ALKALINE R-GROUP: ARGININE (C_6), LYSINE (C_6), HISTIDINE (C_6)

IMINO ACID: PROLINE (C_5)

PEPTIDE STRUCTURE

Fig. 5.1. The amino acids found in proteins.

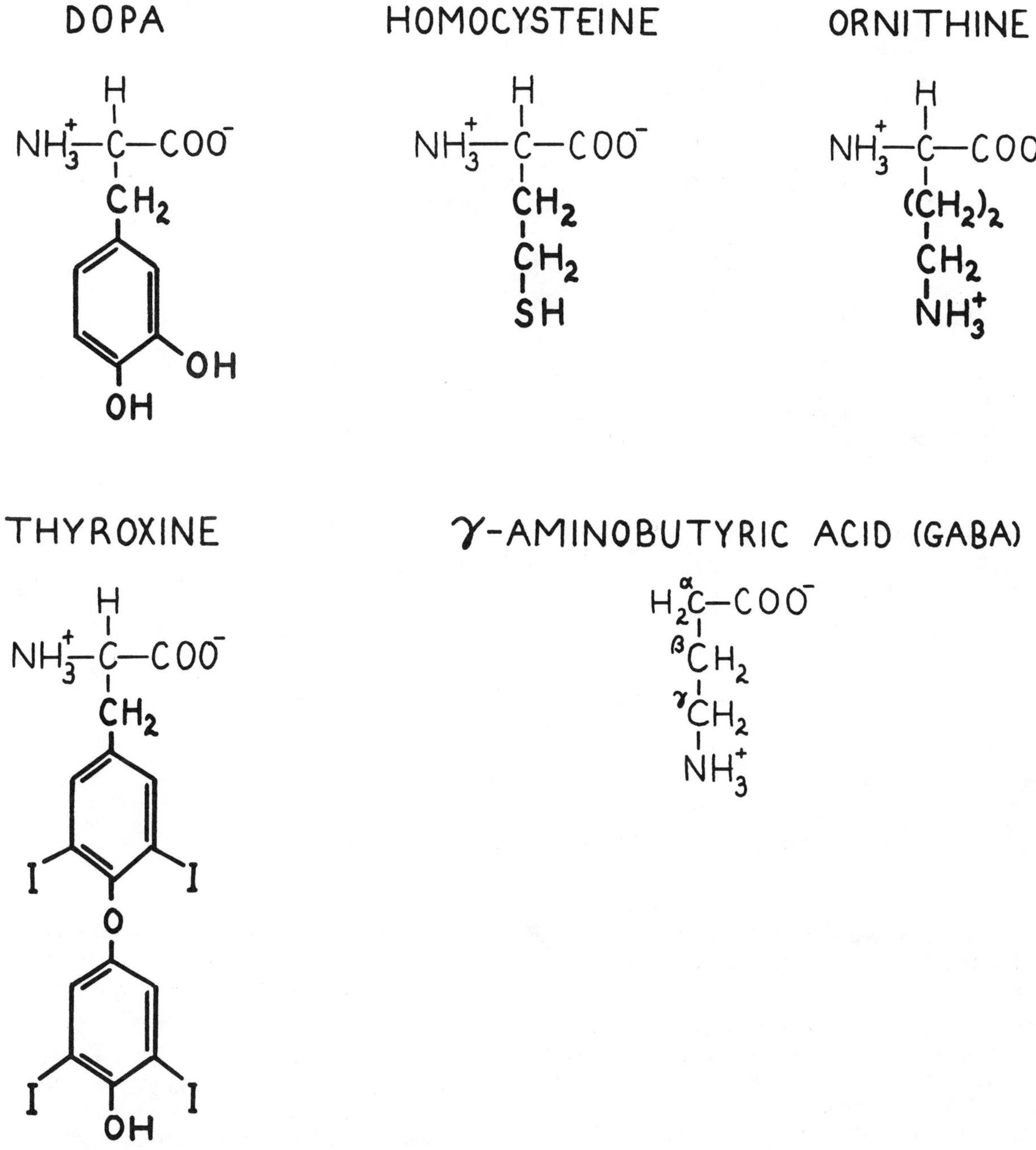

DOPA

HOMOCYSTEINE

ORNITHINE

THYROXINE

γ-AMINOBUTYRIC ACID (GABA)

Fig. 5.2. Examples of amino acids not found in proteins.

do! Figure 5.3 shows a mnemonic for remembering the 10 essential amino acids:

Imagine a **HILL** (**H**istidine, **I**soleucine, **L**eucine, **L**ysine). On one side of the hill is **ME** (methionine) and on the other side, my friend **VAL** (Valine). In order to save the world I have to throw over the hill this Fiend (Phenylalanine—a big fiend at that, having the longest name of the amino acids). If I miss the throw, the world perishes. I have **3 TRIES** (Threonine, Tryptophan). I tried 3 times but failed. Too bad, I missed. Note Val's reaction ("Argghh!"—Arginine). Val is repulsed by the

Fig. 5.3. Mnemonic for remembering the 10 essential amino acids (see text for description).

prospect of catching this fiend. If you fix the picture in mind, it will be difficult to forget the 10 essential amino acids.

"Non-essential" amino acids are those that can be synthesized by the body. They are largely spun off from the Main Powerhouse. For instance, pyruvate becomes alanine (E-7); oxaloacetate becomes aspartate (D-9); 2-ketoglutarate (alpha-ketoglutarate) becomes glutamate (E-11). Amino acids may also be formed from other amino acids. Examples: serine becomes glycine and cysteine (E-7); glutamate becomes glutamine and proline (F-11); aspartate becomes asparagine (C-8); phenylalanine becomes tyrosine (C-7).

Functions of the amino acids

Amino acids may be used as fuel, or converted to other kinds of molecules:

1. **Fuel.** From their vantage point in the Midway, amino acids can generate fuel molecules by entering the Main Powerhouse (e.g., transformation to pyruvate, acetyl CoA, fumarate, succinate, 2-ketoglutarate, or acetoacetate.

Amino acids are glucogenic and/or ketogenic. **Glucogenic** amino acids can convert to glucose. They include those that transform to pyruvate or any of the Krebs cycle intermediates, that, in turn, can convert to glucose. **Ketogenic** amino acids are those that convert to acetyl CoA (or acetoacetate). The latter molecules cannot be converted to glucose. They can, via acetyl CoA, enter the Krebs cycle to generate energy while changing to CO_2 and H_2O but, as the pyruvate-to-acetyl CoA step is one-way, they cannot revert to glucose.

2. Conversion to other molecules. Apart from conversion to fuel molecules, the amino acids are building blocks of other molecules. E.g., aspartate and glutamine contribute to the formation of purines and pyrimidines (fig. 5.4),

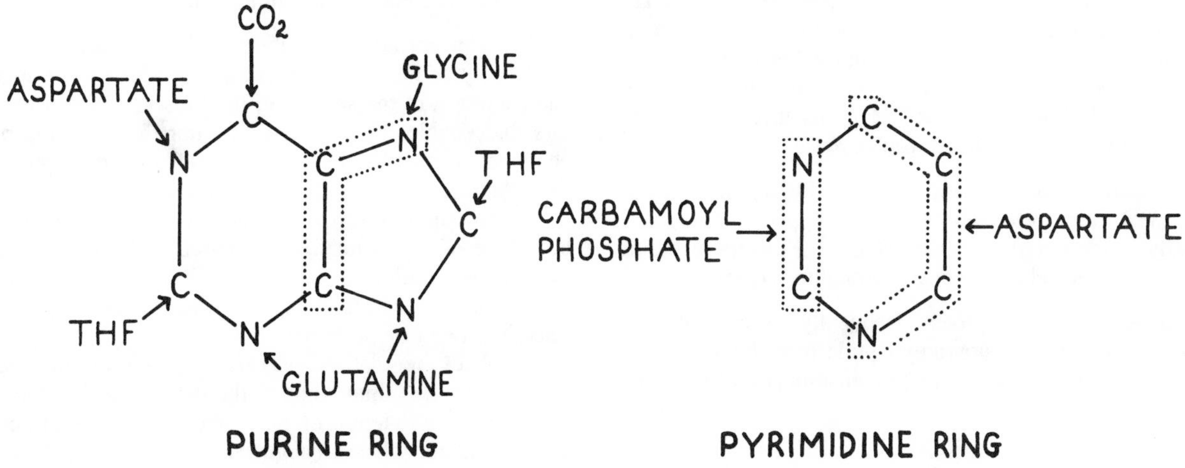

Fig. 5.4. Contributions to the purine and pyrimidine rings. THF, tetrahydrofolate.

which are building blocks of nucleic acids. Serine provides the backbone for the sphingolipid molecules in Sphingo's Curio shop (Fig. 4.5). Histidine becomes histamine (F-11). Tyrosine (C-7) becomes thyroxine, melanin, dopamine and epinephrine. Tryptophan (I-9) becomes serotonin (5-OH tryptamine) and contributes to the nicotinamide ring in NAD^+. Glycine forms part of the porphyrin ring (D-11).

Amino acids, by the way, also form peptides and proteins. A **peptide** is a short chain of amino acids (fig. 5.1) and includes many hormones and other kinds of molecules.

Proteins are very long chains of amino acids with different functions that depend not only on their specific sequences of amino acids (= **primary structure**), but the overall shape of the sequence. This includes its **secondary structure** (for example, forming a spiral molecule), **tertiary structure** (which depends on widely separated portions of the molecule attaching to another, as by sulfur groups, and even **quaternary structure,** which depends on the apposition of different molecules (e.g., collagen molecules combined in several intertwining geometric strands. There are thousands of important proteins, the description of which is beyond the scope of this book. In general, though, proteins may be grouped into certain general functions, as follows:

1. **Transport**—proteins may act as carriers to transport other molecules. For instance, hemoglobin is a protein that carries O_2 in red blood cells; myoglobin transports O_2 in muscle cells; transferrin transports iron; thyroglobulin-binding protein transports thyroxin. Albumin forms the largest proportion of plasma protein. It carries various hormones, iron, heme, vitamins, bilirubin, free fatty acids, Ca^{++}, rare metals and many drugs, enabling them to be soluble in plasma. Attachments to albumin may render many molecules inactive. Albumin also has an osmotic effect that helps to maintain the blood volume. Albumin may be used on an emergency basis, for instance, as a replacement for lost blood.
2. **Storage.** Ferritin, for instance, is a protein that stores iron in the liver.
3. **Motion.** Muscle contraction occurs when filaments containing the proteins actin and myosin slide along one another. In cells with cilia or flagellae, microtubules, which contain tubulin protein, slide along each other, facilitating movement.
4. **Structure. Collagen** is a widespread protein that provides a structural framework of intercellular tissue support in connective tissue, cartilage, bone and other tissues. **Elastin** is a stretchable support protein. **Keratin** is the tough protein of fingernails and hair.
5. **Control of gene expression.** Select proteins act on DNA to control the expression of genes.
6. **Growth substances.** Certain proteins may promote growth and regeneration of tissues in the embryo and adult.
7. **Immune mechanisms.** Antibodies are proteins that help control the spread of infections and eliminate foreign material.
8. **Clotting mechanisms.** The clotting factors, e.g., fibrinogen and thrombin, are proteins.
9. Important components of **cell membranes,** including receptors on cell surfaces that control the passage and/or action of various chemicals that influence the cell.
10. **Hormones.** Many of the hormones are peptides or proteins (e.g., insulin, growth hormone, adrenocorticotrophic hormone). Hormones directly or indirectly influence the degree of synthesis or activation of enzymes.

11. **Enzymes.** Enzymes are proteins that alter the rates of chemical reactions, as opposed to substrates, which are transformed by the reaction. Some of the kinds of general enzymes and reactions that occur in the body are summarized in Appendix II. Note that in general, the name of an enzyme is followed by "ase". For instance, glucose-6-phosphatase acts on glucose-6-phosphate.

Globular proteins (e.g., albumin, hemoglobin, most enzymes) are relatively "globe"-shaped; they have greatly folded and compact chains. **Fibrous** proteins (e.g., collagen, elastin) contain straighter chains.

The function of a protein frequently depends on the attachment to it of a non-polypeptide (prosthetic) group. An **apoprotein** is the active protein minus its prosthetic group. Thus:

apoprotein + prosthetic group = functioning protein.

For example, globin (apoprotein) + heme (a prosthetic group that contains iron in a porphyrin ring) = hemoglobin. In a slightly different useage, apoprotein in "association" with lipid = lipoprotein vehicles that transfer cholesterol and triglycerides in the blood stream.

A **coenzyme** is the prosthetic group of an enzyme. Thus:

apoenzyme + coenzyme = functioning enzyme.

The water-soluble vitamins are coenzymes of a number of enzymes in Biochemistryland.

A **zymogen** is an inactive **precursor** of an active enzyme. Activity may be induced by cleavage of certain bonds in the zymogen or by addition of phosphoryl or other groups. Storing enzymes in inactive form is a good thing. In pancreatitis, for example, normally inactive lipolytic and proteolytic zymogens become activated and eat away at the pancreas. Examples of zymogens include those for digestion and blood clotting. For example, in digestion, the zymogen trypsinogen converts to the active enzyme trypsin; pepsinogen converts to pepsin. In the long chemical cascade of blood clotting, the zymogen prothrombin becomes the active thrombin, which catalyzes the change of fibrinogen to fibrin. Proteins that are not enzymes may also, in some cases, exist in an active and inactive form (e.g., proinsulin becomes insulin; procollagen becomes collagen).

Enzymes are important in preventing the reactions in Biochemistryland from running out of control. They regulate the **speed** of chemical reactions. As mentioned, the **direction** of a chemical reaction is regulated by the energy kinetics of the reaction and by the concentration of substrates. The higher the concentration of substrates and the greater the energy loss in the reaction the greater the tendency for the reaction to occur. A number of factors affect the efficiency at which enzymes act:

1. The concentration of substrate. This depends on the availability of the substrate in the diet, the rate of uptake and excretion of the substrate, the presence of cell receptors that enable the substrate to enter the cell, and biochemical mechanisms within the cell that may produce the substrate.
2. The amount of enzyme produced in the cell. The metabolic rate of a cell may be increased with increased enzyme concentration.
3. The state of the enzyme, i.e., whether it is a zymogen (inactive form) or in its active state.
4. Feedback regulation. Excess reaction products can feedback to the initial steps of the reaction and reduce or increase the efficiency of particular enzymes that act at earlier reaction phases (**negative** or **positive feedback**). For example, excess dietary cholesterol can reduce cholesterol synthesis in the body by inhibiting hydroxymethyl glutaryl CoA reductase (at the step between 3-hydroxy-3-methylglutaryl CoA and mevalonate) (G-10). Many drugs and toxins can also inactivate enzymes. An enzyme's activity may on the other hand be enhanced by certain molecules (positive feedback). For example, AMP stimulates the activity of phosphofructokinase (in the change of fructose 6-P to fructose 1,6-P_2) (D-3). Sometimes feedback can change the actual specificity of an enzyme. For example, lactose synthase catalyzes the combination of UDP-galactose with protein to form a glycoprotein. In pregnancy, though, lactose synthase is modified to form an enzyme complex that instead facilitates the conversion of UDP-galactose to lactose, a prime ingredient in milk. The activity of various enzymes can be modified across far distances through the effects of hormones, which in indirect ways may increase or decrease an enzyme's level of functioning.

A number of reactions in the body involve multiple cascading steps in which zymogens are successively transformed into active enzymes. A classic example is the long chemical cascade in Carbohydrateland, in which adenylate cyclase activation eventually results in phosphorylase activation (fig. 3.2). The sequence of reactions that eventually leads to blood clotting following trauma is another example of such a cascade.

One may inquire as to the value of such cascades. Why not use a single step instead of multiple steps with multiple enzymes? There are at least two advantages to a cascade:

1. It provides a way in which a very small amount of enzyme can produce a large and rapid response through a snowballing reaction.
2. The cascade provides multiple places in which feedback reactions can alter the chain. It is unreasonable to expect that a multitude of different reaction products would coincidentally be able to feedback to alter the activ-

ity of a single enzyme. With multiple kinds of enzymes in a cascade, though, there is much greater opportunity for feedback from diverse sources.

Proteins may form protein or glycoprotein **cell surface receptors.** Specific molecules that contact such receptors can trigger a sequence of reactions within the cell.

The Urea Rest Room (A-9)

There is a Urea Rest Room in the Amino Acid Midway where amino acids are degraded and excreted:

The Amino Acid Midway does not have a storage area as do Carbohydrateland and Lipidland. Excess amino acids are used as fuel or converted into other kinds of molecules. When transformed to other molecules, amino acids commonly give up their $-NH_2$ groups, which are excreted in the form of urea in the Urea Rest Room. The remainder of the amino acid molecule (now minus its $-NH_2$ group) is then free to become some other molecule, whether it be a Krebs cycle intermediate or something else. For instance, amino acids can be converted into Main Powerhouse molecules. They first must lose their $-NH_2$ groups, as the Main Powerhouse molecules do not contain $-NH_2$ groups. The amino acids lose this ammonia in the urea rest room. In a sense amino acids must enter the rest room before entering the pool of powerhouse chemicals just as one must shower before entering a public pool. There is a bouncer in the powerhouse named 2-ketoglutarate (alpha-ketoglutarate) which will physically remove ammonia from the amino acid and deposit the ammonia in the rest room. In grabbing the ammonia, 2-ketoglutarate becomes glutamate, which has a special talent for depositing ammonias in the rest room, and in doing so, can transform itself back to 2-ketoglutarate. Thus, the 2-ketoglutarate can be replenished. The bouncer's job is called **transamination:** the 2-ketoglutarate (E-11) transfers one of its oxygens to the amino acid while removing and taking on the amino acid's $-NH_2$ group to form glutamate. Glutamate may then release ammonia (**oxidative deamination**) to the urea cycle, and in the process change back to 2-ketoglutarate (E-11).

Both glutamate and aspartate are quite proficient in depositing their $-NH_2$ groups in the urea cycle. Aspartate, in fact is a part of the urea cycle. Glutamate and aspartate are so proficient at unloading $-NH_2$ that they are also called upon to donate $-NH_2$ groups in the formation of purines, pyrimidines, amino sugars, and other molecules. Both purines and pyrimidines contain nitrogens in their rings; glutamate and aspartate are heavy contributors of these nitrogens (fig. 5.4).

Gamma amino butyric acid (GABA) is included in this chart (F-11) as it is an important inhibitory neural transmitter.

Where Is the Amino Acid Midway?

Synthesis of non-essential amino acids and proteins can occur throughout the body. Formation of urea occurs predominantly in the liver. The urea is then excreted mainly in the urine (and sweat). Special derivatives of amino acids originate in specific organs. For instance, thyroxine originates in the thyroid gland. Norepinephrine originates in cells of the sympathetic nervous system. Dopamine is present in a number of neuronal tissues, where it can act as a neural transmitter. Melanin originates in the pigmented cells of the skin and in two areas of the brain—the substantia nigra and the locus coeruleus.

Breakdown of ingested proteins occurs in the gastrointestinal tract by digestive enzymes: pepsin from the stomach; trypsin, chymotrypsin, and carboxypeptidase from the pancreas; aminopeptidases and dipeptidases from the small intestinal wall. (Dipeptidases work on dipeptides. Aminopeptidase attacks the amino end of a polypeptide chain whereas carboxypeptidase attacks the carboxyl end of a polypeptide). Such digestion is certainly vital as 10 of the amino acids are essential, having to be acquired in the diet.

Summary Of Connections Of the Amino Acid Midway

1. Ketogenic and glucogenic amino acids can convert to **Main Powerhouse** molecules. Nonessential amino acids can be produced directly or indirectly from Main Powerhouse molecules.
2. Amino acids can combine with molecules from **Carbohydrateland** to form amino sugars and glycoproteins. They can also combine with lipids from **Lipidland** to form lipoproteins. These will be further examined in Combo Circle.
3. The Amino Acid Midway contains a **rest room,** where amino acids are stripped of ammonia, which is then excreted as urea.
4. The Amino Acid Midway connects with the **DNA Funhouse,** in the formation of purines and pyrimidines.

CHAPTER 6. COMBO CIRCLE

The different elements of Carbohydrateland, Lipidland, and the Amino Acid Midway can combine in different ways to form lipoproteins, glycolipids, and glycoproteins. These are the three theatre sections of Combo Circle (G-1).

Liprotein Theatre

Lipoproteins are best known as the trams that help transport triglycerides and cholesterol throughout the bloodstream. The latter lipids are otherwise not very soluble in water. Lipoproteins have both polar and non-polar groups. They can solubilize lipids internally through their non-polar groups and simultaneously be soluble in the bloodstream through their polar ends. The walls of the lipoprotein trams are partly composed of cholesterol, phosphopholipids, and apoproteins (fig. 6.1). There are many apoprotein types, and the particular type helps determine the specific function of the lipoprotein. Triglycerides and cholesterol esters are solublized within the lipoprotein vehicles.

Major lipoprotein types include chylomicrons, VLDL (very low density lipoproteins), LDL (low density lipoproteins), and HDL (high density lipoproteins). This categorization is a reflection of different lipoprotein weights, as may best be seen on ultracentrifugation (fig. 6.2). Lipoproteins also can be separated using electrophoresis (fig. 6.2). Electrophoresis separates serum into a number of protein-containing bands (albumin, alpha 1 and alpha 2 globulin, beta globulin, and gamma globulin, respectively, on moving closer to the origin (cathode). When the pattern is stained for **lipids,** HDL is found with the alpha 1 band; LDL with the beta band; VLDL as a pre-beta band; and chlomicrons at the origin. Ultracentrifugation is good at quantitizing chylomicron, VLDL, LDL, and HDL fractions. Electrophoresis is good at analyzing variations in apoprotein content that become manifest as charge differences. Both ultracentrifugation and electrophoresis may be used to analyze clinical disorders of the lipoproteins.

In order to better understand the individual lipoprotein functions, let us go through an overview of lipid digestion, absorption and distribution:

Say you eat a hamburger. Included in hamburgers are triglycerides and cholesterol esters. (A cholesterol ester is cholesterol in which a fatty acid is attached to cholesterol's −OH group). As neither the triglycerides nor the cholesterol esters have any free polar −OH groups they are not very soluble in water. They are a fatty lump in the stomach. They must be digested, absorbed and distributed in the body:

Digestion

1. The stomach mechanically breaks up fat into smaller particles.
2. **Bile salts,** acting as **micelles** (fig. 6.3), emulsify the fats in the small intestine, breaking them up into tiny soluble droplets that provide a broad surface area for interaction with digestive enzymes.
3. **Pancreatic lipase** facilitates the breakdown of triglycerides into fatty acids and monoglycerides. (For some

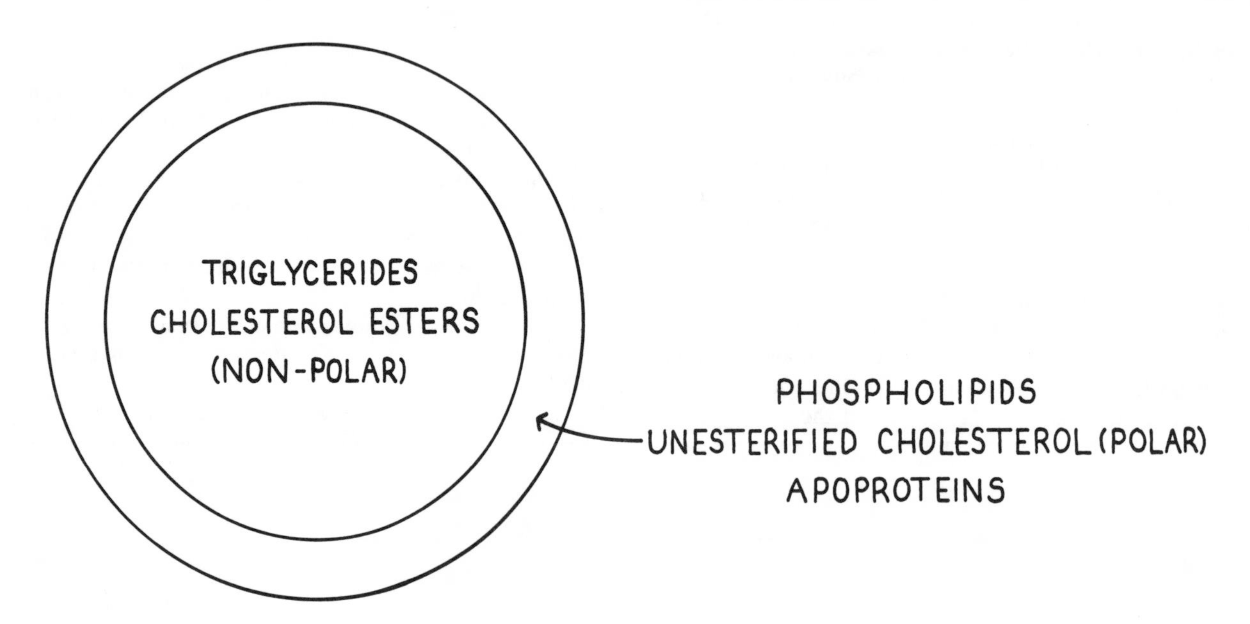

Fig. 6.1. The structure of lipoprotein trams.

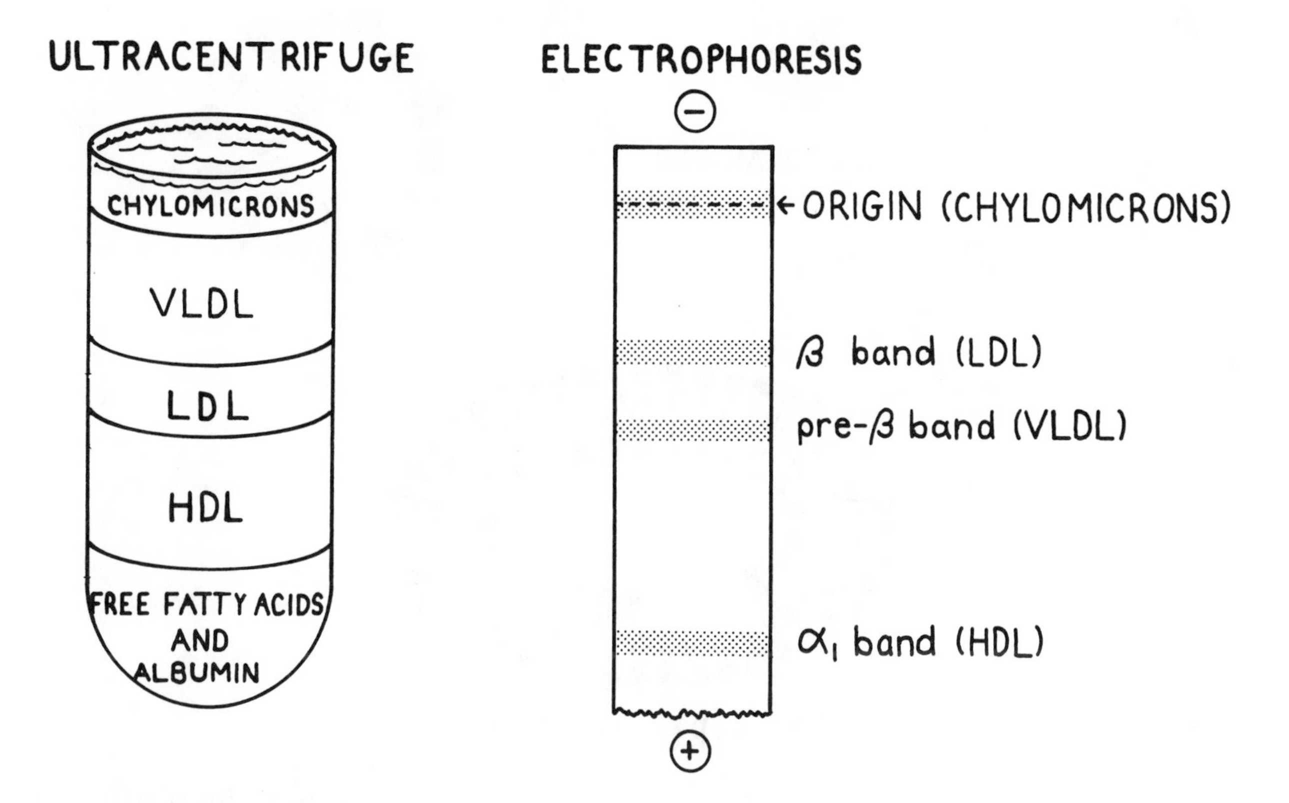

Fig. 6.2. Separation of lipoproteins by ultracentrifugation and electrophoresis.

reason, the center fatty acid of the triglyceride is left alone, resulting in a monoglyceride. Perhaps the center fatty acid does not taste so good to the lipase).
4. **Pancreatic and intestinal wall esterases** break down cholesterol esters to free cholesterol and fatty acids.

Absorption

1. The free fatty acids, monoglycerides, and free cholesterol enter the small intestinal cell.
2. The monoglycerides and free cholesterol are largely reesterified (with free fatty acids) in the intestinal cell. The resultant triglycerides and cholesterol esters (some free cholesterol, too) are incorporated into chylomicrons which carry the triglycerides and cholesterol in the **lymph** stream to the general circulation.
3. Some of the free fatty acids (the small and medium chain fatty acids) are not incorporated into chylomicrons but enter the **venous** circulation directly. They are not too soluble, however, and travel attached to albumin.

Lipoprotein distribution (fig. 6.4)

1. **Chylomicrons.** Where would you go if you were a chylomicron? For one thing, triglycerides, the main component of chylomicrons, are stored in fat cells, so chylomicrons go there. Also, fatty acids from triglycerides are used as energy sources, particularly in muscle cells. The chylomicrons drop off their triglycerides largely in fat and skeletal and heart muscle cells. A **lipoprotein lipase,** on the capillary endothelial cells of the muscle and fat tissues, first releases free fatty acids from the triglycerides. The free fatty acids then enter the muscle or fat cells, where they may be reesterified to triglycerides. The remaining "remnant" particles, which are now rich in cholesterol, go to the liver. They go there because everyone in Biochemistryland knows that the liver knows what to do with cholesterol. The liver has a lot of experience with cholesterol, being in fact, a primary synthesizer of cholesterol (as well as triglycerides). When presented with an excess of cholesterol, the liver can excrete it as such or in the form of bile salts, or it can export it elsewhere in the body, along with triglycerides, as VLDL particles.
2. **VLDL, IDL and LDL.** VLDL particles are released by the liver and are rich in triglycerides and cholesterol. VLDL particles, like the chylomicrons, drop off their triglycerides in fat and muscle cells. The remainder of the particles become cholesterol rich LDL particles (after going through an intermediate IDL stage). LDL distributes the cholesterol widely to hepatic and non-hepatic cells

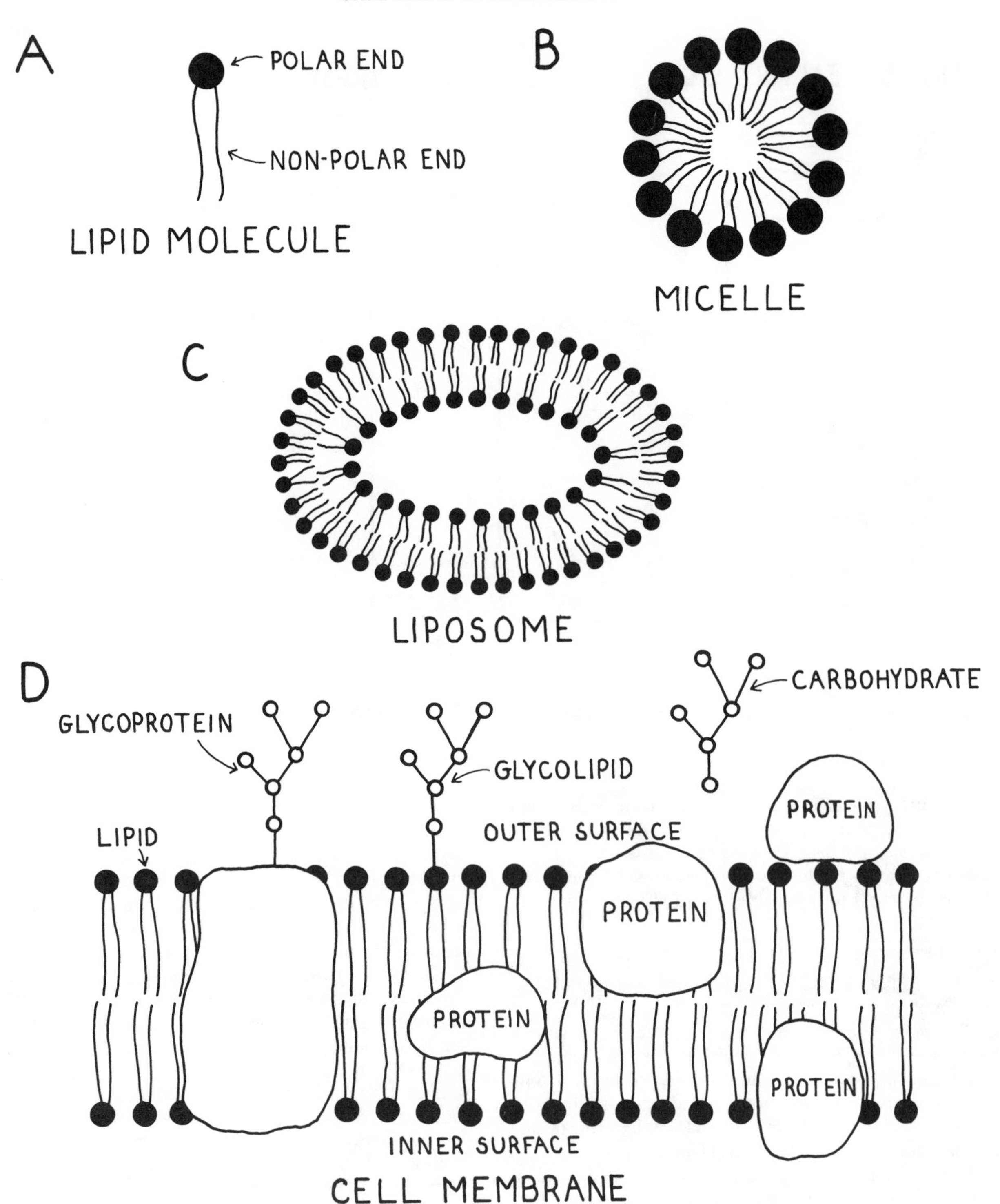

Fig. 6.3. Micelles, liposomes and cell membranes. **Micelles** are collections of lipid molecules that are relatively non-polar internally and polar externally. This arrangement allows relative water-solubility of the micelle as a whole. **Liposomes** contain lipid molecules in a bilayer. They may be used as artificial vehicles for trapping and delivery of drugs to specific tissues. They are also useful as models of cell surface function. A real cell membrane is not only a lipid bilayer, but also includes proteins, glycoproteins, glycolipids, and lipoproteins (lipoproteins actually are a mixture of lipids and proteins rather than fused lipid-protein molecules). The "glyco" attachments on the outer surface may be important in labeling cells with specific cell-surface properties.

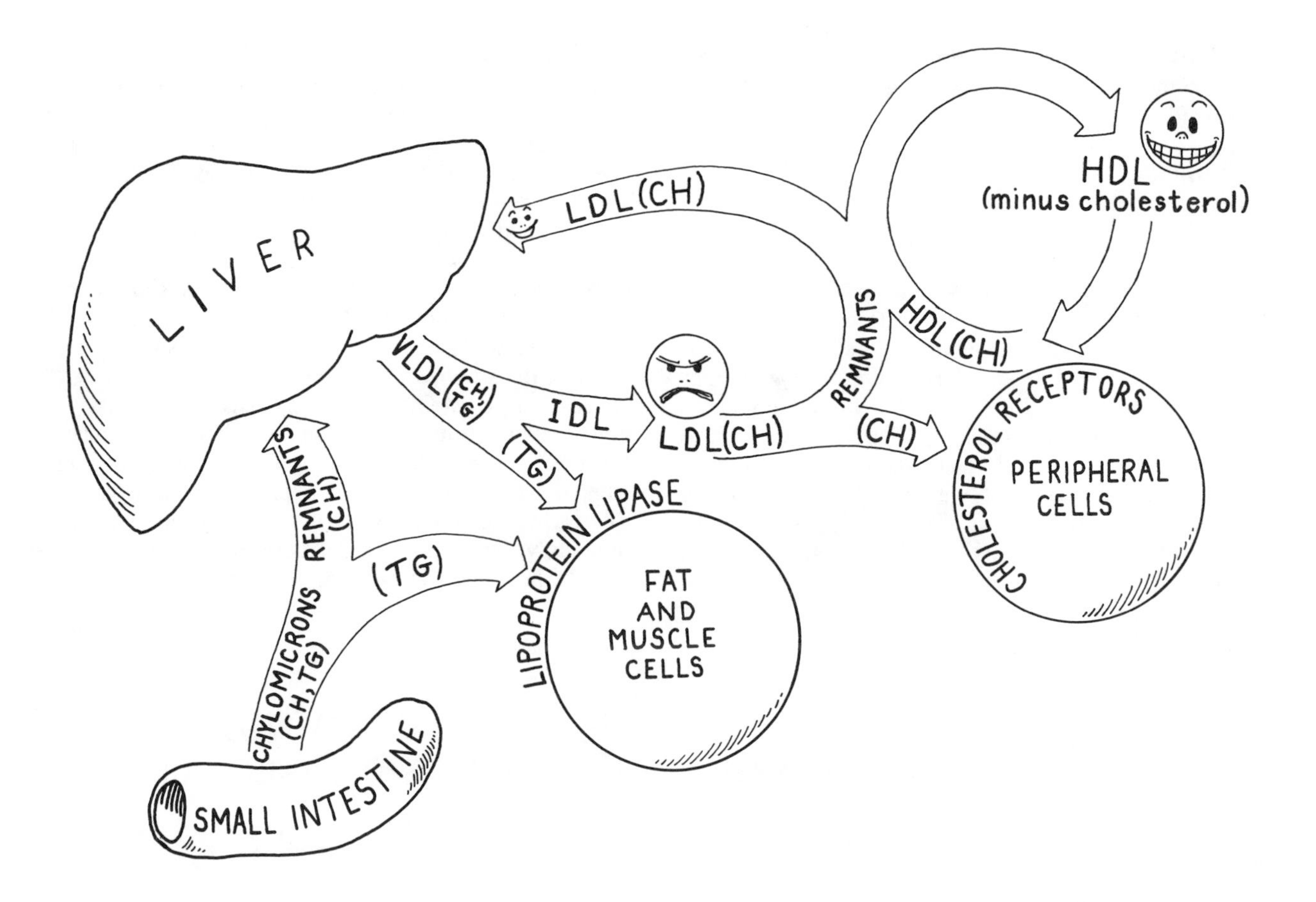

Fig. 6.4. The distribution of lipoproteins in the body (see pg. 56 for clinical discussion).

throughout the body. This is a good idea, as cholesterol is not only useful in the formation of specialized things like bile salts and steroid hormones, but is a major component of cell membranes in general.

3. **HDL and LCAT.** While all this is going on, there is a scavenger called HDL which carries unwanted, excess cholesterol, partly from cell breakdown, back to the liver (largely within LDL remnants) where the cholesterol might end up being excreted (for instance, as bile salts). LCAT (lecithin-cholesterol acyl transferase) is an enzyme associated with HDL that reesterifies free cholesterol.

Glycolipid Theatre

Glycolipids (fig. 4.5) are combinations of sugars and lipids. The backbone of the glycolipid molecule is **ceramide,** which was mentioned in the section on Sphingo's Curio Shop. To review, **serine** is a 3-carbon coat rack. When a fatty acid is hung on its −COOH group, it becomes **sphingosine.** When a further fatty acid attaches to its −NH2 group, it becomes **ceramide.** This leaves the ceramide's third carbon with its −OH group free for reaction with other molecules. If phosphocholine attaches here, we get **sphingomyelin.** If **sugars** attach, we get **glycolipids,** which include **cerebrosides** and **gangliosides,** important components of cell membranes, expecially in the nervous system.

Cerebrosides are simpler than gangliosides. Only one monosaccharide attaches for cerebrosides, and that generally is either glucose or galactose.

Sulfatides are cerebrosides that also contain a sulfate group attached to the galactose part of the molecule. Sulfatides are found in significant amounts in the brain.

If longer, branched chain sugars attach to the third carbon −OH group, the molecule is more complex and is called a **ganglioside** (the sugars "gang up" in a ganglioside). In addition to glucose and galactose, gangliosides contain at least one molecule of a sialic acid (an 11-carbon amino sugar, also called N-acetyl neuraminic acid, or NANA), and may contain other sugar groups as well. The sugars are combined in complex branched ways and attach to the ceramide, generally via a glucose residue.

Globosides are glycolipids that are intermediate in complexity between the cerebrosides and gangliosides. Namely, they have two or more sugar residues, but no sialic acid.

Glycoprotein Theatre

This theatre contains sugars that combine with amino acids, as well as sugars that combine with polypeptides and proteins.

A. **Sugars that combine with amino acids.** An initial first step in the formation of amino sugars is the combination of fructose 6-P with an −NH2 group from glutamine to form the 6-carbon glucosamine 6-P (an amino sugar) (E-2). Subsequent transformations lead to other amino sugars, such as N-acetyl galactosamine, an 8-carbon amino sugar that is commonly a component of globoside and ganglioside molecules. (It also just happens to be the specific portion of the (antigen) molecule that determines blood group type A, whereas galactose determines blood type B). **Sialic acids** are 11 carbon amino sugars. They are an important component of gangliosides.

B. **Sugars that combine with peptides and proteins.** Glycoproteins are important components of cell membranes. Many of the antigenic sites and hormone receptor sites on cells are glycoproteins that provide the cell with a specific identity tag. The antigens associated with the A,B,O blood groups are glycoproteins. So are many other proteins in the blood, including the immunoglobulins, complement, and blood clotting proteins. Certain hormones are glycoproteins (TSH and FSH). Interferon, an antiviral agent, is a glycoprotein. Collagen is in part associated with carbohydrates that attach to collagen hydroxylysine groups. Many secretions contain a subgroup of glycoproteins **(mucins)** which may have a lubricative function. When carbohydrates overwhelmingly predominate, the molecule is called a **proteoglycan (fig. 6.5).** The carbohydrate portion of the proteoglycan is called a **mucopolysaccharide** (or **glycosaminoglycans**). Typically, a proteoglycan contains a central protein core, to

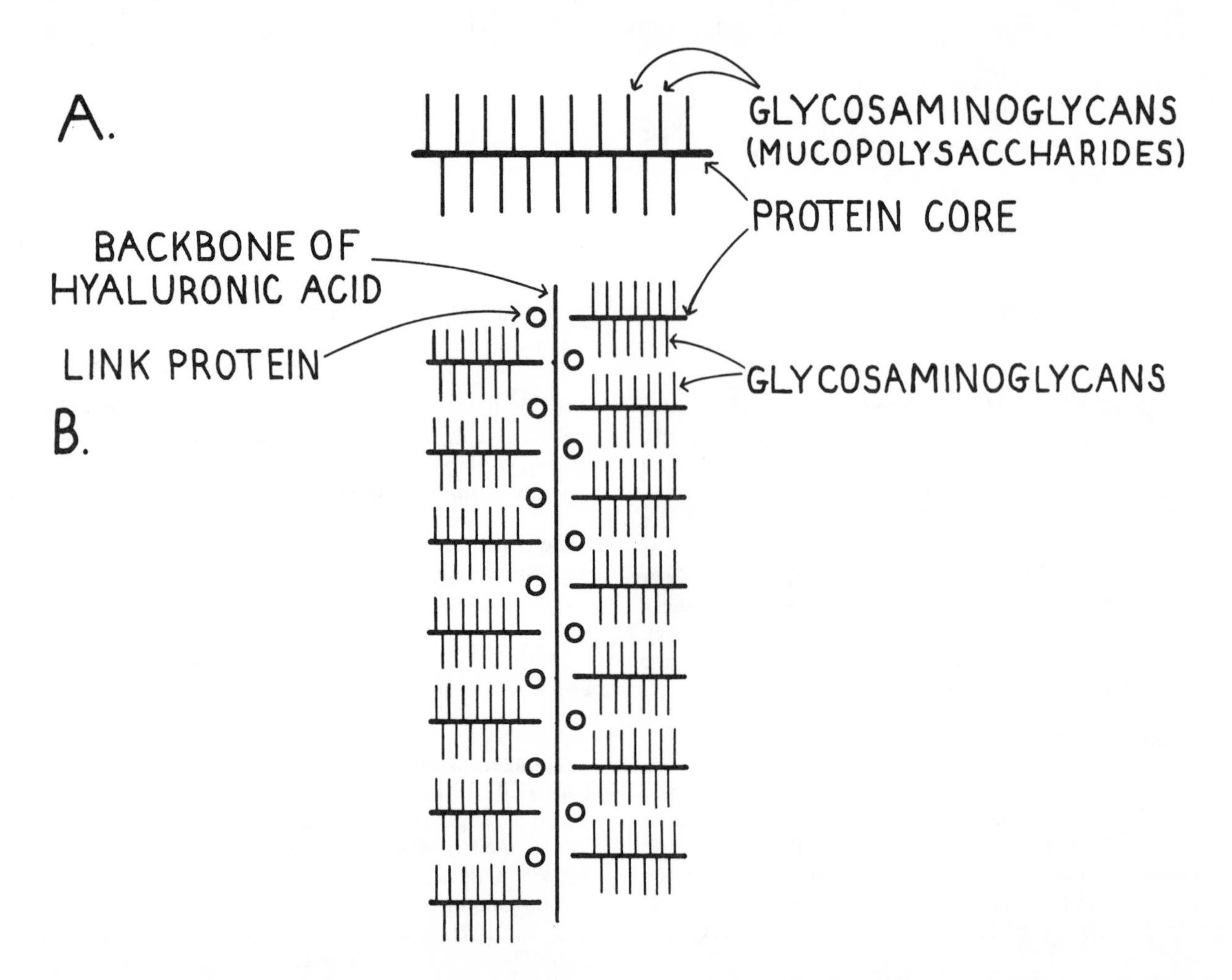

Fig. 6.5. A proteoglycan molecule contains a protein core to which numerous carbohydrate (glycosaminoglycan) groups attach. Proteoglycan units can aggregate along a central hyaluronic acid backbone.

which are attached numerous mucopolysaccharide chains, commonly including hexosamine sugars and uronic acid (carboxyl-containing) sugars, sometimes with sulfate groups attached to the sugars. Some of the common mucopolysaccharides include:

Hyaluronic acid
Chondroitin sulfate
Dermatan sulfate
Heparin
Heparan sulfate
Keratan sulfate

Mucopolysaccharides absorb significant quantities of water and tend to have a mucoid, or viscous consistency. This enables them to provide a structural or lubricatory role in connective tissue. They also help to maintain fluid and electrolyte balance throughout the body. **Hyaluronic acid** differs from the other five types in that it has not been shown to attach along the core protein, but does provide a backbone for proteoglycan aggregates (fig. 6.5). Also, it is not sulfated.

Glycogen is believed to be a proteoglycan, as glycogen chains appear to require a backbone primer protein on which the synthesis of glycogen chains is initiated. The final product is a large proteoglycan molecule that contains numerous glycogen chains attached to the protein primer backbone.

Where is Combo Circle?

The **Golgi apparatus** and **lysosomes** are important intracellular organelles that relate to the molecules of Combo Circle. The Golgi apparatus is important in the modification of existing proteins and in the construction and packaging of complex macromolecules. Lysosomes are important in the degradation of complex macromolecules.

The lipoproteins shuttle in the blood stream between liver, fat, muscle, and other peripheral cells. Chylomicrons originate at the level of the intestinal cell; VLDL originates from the liver. The glycoproteins and glycolipids, being important components of cell membranes, are widespread in their origin. Cerebrosides and ganglioside glycolipids are particularly important in neuronal cell membranes.

And now it's time for the DNA Funhouse!

CHAPTER 7. THE DNA FUNHOUSE

Purines and Pyrimidines (fig. 7.1), are best known as building blocks of RNA (ribonucleic acid) and DNA (deoxyribonucleic acid). Adenine and guanine are purine bases (so are xanthine and hypoxanthine, but these are reaction intermediates and are not part of the DNA or RNA molecule). Cytosine, uracil, and thymine are pyrimidine bases.

Note in figure 7.1 that the pyrimidines look something like purines except that they are **not so pure** any more, having been **CUT** (**C**ytosine, **U**racil, **T**hymine) in half, leaving a single hexagonal ring rather than the combined pure hexagonal-pentagonal structure of the purines. Thus, purines are **"pure and unCUT"**.

DNA and RNA each contain purine and pyrimidine bases plus sugar and phosphate groups. Note the following terminology:

Purine or pyrimidine + sugar (ribose or deoxyribose) moiety = a nucleoside
nucleoside + phosphate moiety = a nucleotide
Nucleotides strung in sequence = DNA or RNA

The DNA Funhouse contains a number of levels, the bases being at the bottom (in the basement), with successively higher levels of nucleosides, nucleotides (mono-, di-, and triphosphates), and, finally, DNA and RNA (multiple nucleotides) (B-6).

Both DNA and RNA contain cytosine, adenine, and guanine. In general, though, thymidine is specific to DNA, whereas uracil is specific to RNA. RNA, in general, is a single strand, whereas DNA is a double strand that contains hydrogen bond cross linkages between purine and pyrimidine residues. Thymidine cross-connects with adenine (by hydrogen bonding), and cytosine cross-connects with guanine (fig. 7.2). DNA and RNA also differ in that the sugar portion of DNA is a deoxyribose, whereas it is simple ribose in RNA (hence, DNA=Deoxyribonucleic acid and RNA=ribonucleic acid). The two strands of the DNA helix are arranged in an antiparallel manner, with the 5′ to 3′ sequence in line with a 3′ to 5′ sequence (fig. 7.2). In order for the base pairs to match each other (joining via hydrogen bonds), the base pairs are situated at the central core of the helix, with the phosphate and sugar groups more peripheral.

The key events that occur in the formation of DNA, RNA and protein are called REPLICATION, TRANSCRIPTION, and TRANSLATION, respectively:

Replication (Reproduction of DNA)

A. **Primase** synthesizes a short strand of RNA (on the DNA template), which acts as a primer in initiating DNA polymerization.

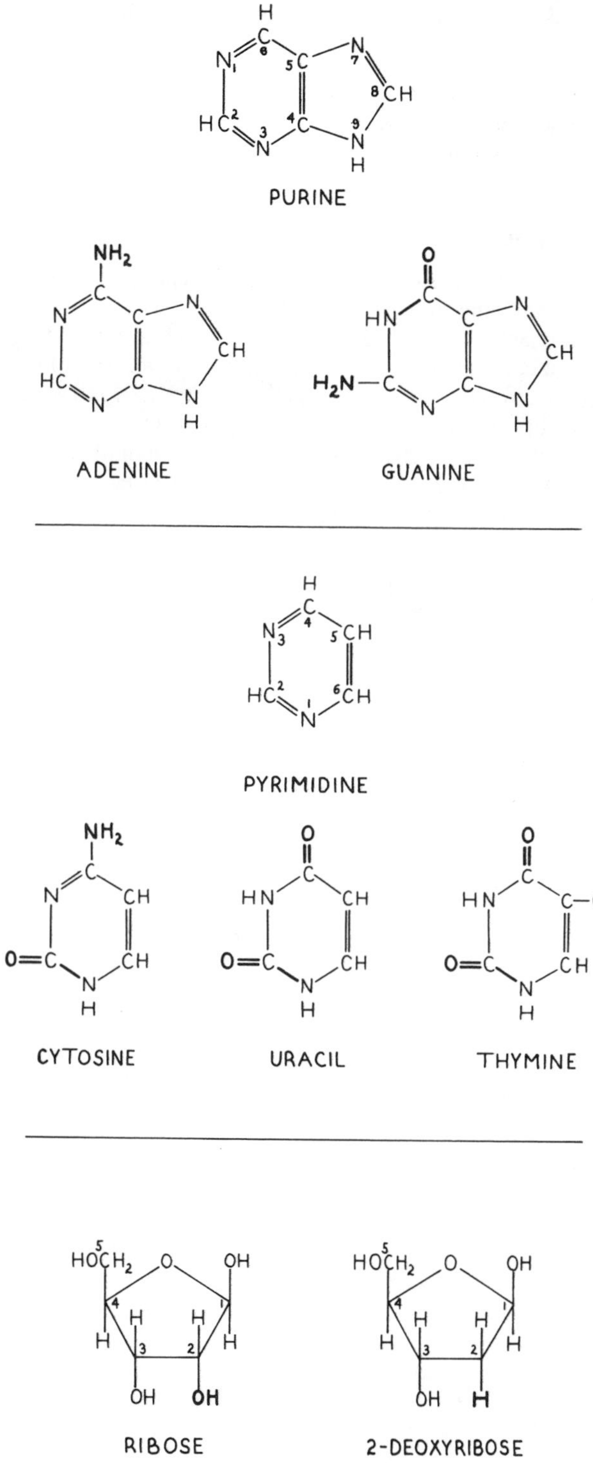

PURINE

ADENINE

GUANINE

PYRIMIDINE

CYTOSINE

URACIL

THYMINE

RIBOSE

2-DEOXYRIBOSE

Fig. 7.1. The basic purine and pyrimidine structure.

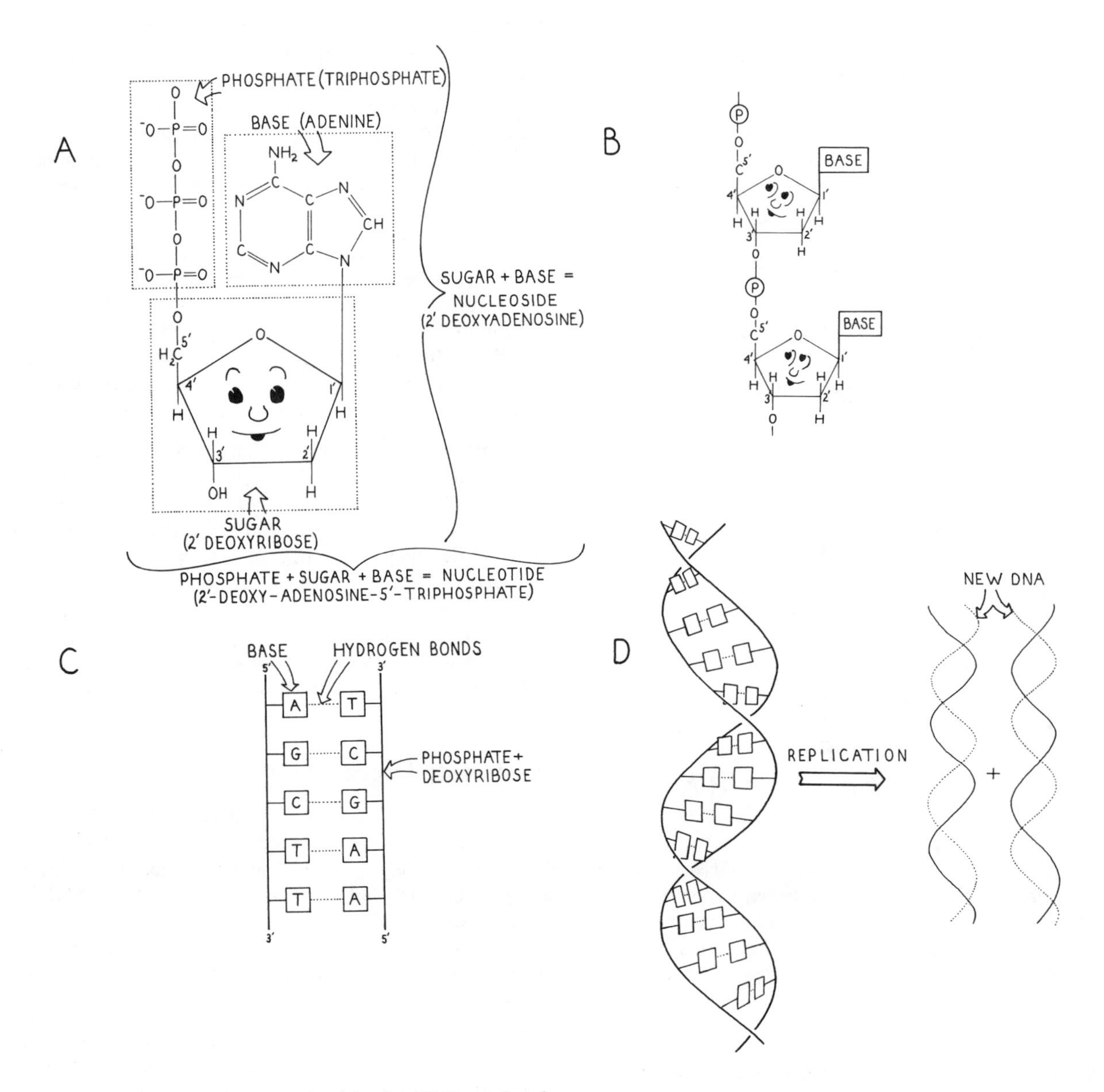

Fig. 7.2. Nucleotide connections in the DNA molecule.

B. **DNA polymerase** sequentially polymerizes nucleotides to form a new DNA strand. DNA polymerases also have **exonuclease** properties—the ability to hydrolyze and remove (or refuse to allow the addition of) improperly paired terminal nucleotides.

C. **DNA ligase** connects newly formed fragments (Okazaki fragments—fig. 7.3) of DNA with one another to form a single chain.

D. **Endonucleases** can hydrolyze connections between those nucleotides that lie within the central area of the nucleotide chain (as opposed to exonucleases, which hydrolyze terminal positions). A specific endonuclease can help remove a segment of DNA rendered defective by ultraviolet light. The latter can damage DNA by making two adjacent thymine bases fuse convalently. The endonuclease makes a nick in the DNA strand on one side of the defective thymine dimer. An exonuclease removes the damaged dimer. A DNA polymerase adds replacement nucleotides. DNA ligase reattaches the newly synthesized DNA to the original strand.

Transcription (Formation of RNA from DNA)

A. **RNA polymerase** sequentially polymerizes nucleotides on the DNA template to form an RNA copy of one of the DNA strands.

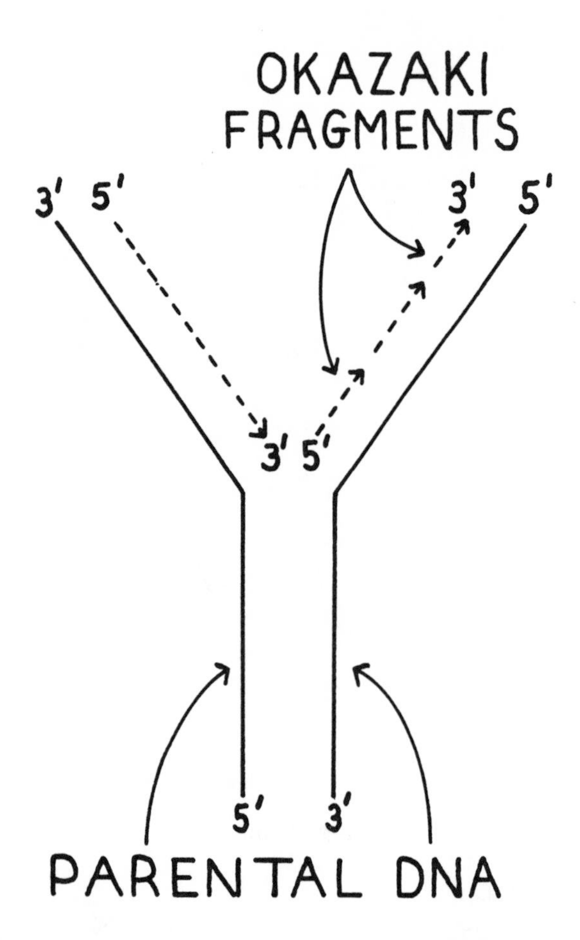

Fig. 7.3. Okazaki fragments. As new DNA must be formed in a 5′ to 3′ direction, this presents little problem for one of the DNA strands (the one on the left), but must be done in short segments (Okazaki fragments) for the other strand. DNA ligase connects the Okazaki fragments.

B. **Promoters** are regions on DNA that signal the RNA polymerase where to begin polymerization. DNA also contains **stop signs** to signal where the RNA transcription should stop. The resulting RNA molecule may subsequently be modified by adding or detracting from the RNA chain. Three general types of RNA result from transcription: **Messenger RNA** carries the primary message of the specific protein to be synthesized on the ribosomes. **Transfer RNA** transfers individual specific amino acids to be linked up on the messenger RNA molecules. **Ribosomal RNA** is an integral part of the ribosomal structure, but its function is unclear.

Translation (Formation of protein on the messenger RNA molecule)

Messenger RNA carries the genetic code in the form of purine and pyrimidine base triplets, called **codons.** Each codon correspond to a specific amino acids on a transfer RNA molecule. An **anticodon** is the recognition site on a transfer RNA molecule that recognizes a specific messenger RNA codon, enabling the transfer RNA, with its attached amino acid, to line up properly on the messenger RNA. It is not the amino acid that attaches to the messenger RNA codon, but the anticodon section of the transfer RNA molecule that attaches.

A given amino acid can match more than one kind of codon. However, a given codon can normally match only one amino acid. There are thus 61 codons to match only 20 amino acids. There are also 3 **terminal** codons that specify a **stop** signal to protein synthesis. **Amino acyl-tRNA synthetase** is an enzyme that attaches specific amino acids to their corresponding transfer RNA.

Beyond translation, proteins and peptides may be further modified by metabolic events within the cell. For instance, the hydroxyproline and hydroxylysine in collagen are formed from proline and lysine that are hydroxylated while they are part of the precursor protein molecule. A protein may also be post-translationally modified by cleavage of select bonds and/or by addition or subtraction of various kinds of groups. Insulin, for instance, is formed by cleavage of a larger proinsulin molecule. Similar events occur in the activation of zymogens.

Nucleotide Degradation

Nucleotides, apart from linking up to form DNA and RNA, may instead be degraded and associated with other kinds of molecules. Adenine and guanine, for instance, are part of ATP and GTP. Adenine is part of NADH, $FADH_2$, CoA, and cyclic AMP. Uracil is part of UDP-glucose, an important intermediate in forming glycogen (D-1). CDP-choline is an intermediate in lecithin synthesis (G-4).

The purine section of the Funhouse has its own restroom in which urate is formed as a waste product of purine degradation. Urate is then excreted in the urine (B-5).

The pyrimidine section of the funhouse does not have its own restroom. The pyrimidines, rather than forming urates, tend to convert to other molecules that can be used in other metabolic reactions. For instance, thymine, may eventually become 3-aminobutyrate (B-6), which eventually may become succinyl CoA, an important molecule in the Krebs cycle. Cytosine and uracil may convert to alanine (B-7).

The pathway to **purine** synthesis is rather long. Parts of the purine molecule are made from glycine, tetrahydrofolate (a one-carbon donor), glutamine, CO_2, and aspartate (fig. 5.4) (B-3). This pathway enters the purine section of the DNA Funhouse at "IMP" (Inosine monophosphate), at the ground level. The body can avoid part of the long synthesis effort by **salvaging** purines that have already been formed and using them to make new nucleotides by the general reaction:

PRPP + purine→ purine nucleotide + PPi

Pyrimidines are synthesized in part from molecules of aspartate and carbamoyl phosphate (fig. 5.4) (B-8). This pathway enters the pyrimidine section of the DNA Funhouse at "UMP" (uridine monophosphate).

Where Is The DNA Funhouse?

As DNA and RNA are common to all cells that have nuclei, the reactions in the Funhouse are present wherever nucleated cells are found. In the gut, nucleic acids are broken down to simpler purines and pyrimidines, but the absorption of purines and pyrimidines is relatively insignificant. The purines and pyrimidines and their derivatives are mainly synthesized within the cells of the body, rather than absorbed in the gut.

Summary of Connections of the DNA Funhouse

1. The DNA Funhouse connects with **Carbohydrateland,** as ribose 5-P is used in forming purines and pyrimidines.
2. The DNA Funhouse connects with the **Amino Acid Midway,** as amino acids are used to form purines and pyrimidines. Also, alanine is a breakdown product of cytosine and uracil.
3. It connects with the Krebs cycle of the **Main Powerhouse** via the **Urea Rest Room,** as aspartate and glutamate, which are spun off the Krebs Cycle are important participants in the Urea Cycle. Also, 3-aminobutyrate, a product of thymine degradation, may connect to succinyl CoA.

CHAPTER 8. PORPHY'S HEMELAND

If you are now ready for blood and guts adventure, let us visit Porphy's Hemeland. Succinyl CoA, a molecule in the Krebs cycle ferris wheel, is the entry point from the Main Powerhouse to Porphy's Hemeland, where one may find the **porphyrin** pinwheels (porphyrin rings) (C-12).

Porphobilinogen contains a single pyrrol ring. It converts to the quadruple-ringed **porphyrinogens** and **porphyrins,** all of which contain four pyrrhol rings arranged like the petals of a pinwheel. One of the porphyrins (protoporphyrin IX) is famous for its combination with Fe^{++} to form heme (D-13). Many important molecules are derivatives of these build-up quadruple pyrrhol petals, including not only heme but such diverse molecules as **vitamin B_{12},** the **cytochromes, catalase, peroxidase,** and (in plants) **chlorophyll.** Chlorophyll resembles heme but, among other things, notably contains a central magnesium instead of an iron atom, which is present in heme. Vitamin B_{12} contains a central cobalt ion. Chlorophyll also contains an attached isoprenoid chain (hence, the location of chlorophyll in Channel No. 5 as well).

The cytochromes are important as electron transporters in the chain of oxidative phosphorylation. Chlorophyll absorbs light energy and converts it to chemical energy: high energy electrons move from the chlorophyll molecules, as they do for the cytochromes, along an electron transport chain that generates ATP and NADPH.

Heme is formed by the combination of protoporphyrin and iron. Hemoglobin is formed by the combination of heme with globin protein. Heme is the prosthetic group for a number of important molecules, including hemoglobin, myoglobin, the cytochromes, and certain enzymes (e.g. catalase, peroxidase). Interestingly, hemoglobin and cytochrome C both have the same heme group but have different associated proteins. This results in their different functions, heme being a carrier of O_2 and cytochrome C being a carrier of electrons. The heme in catalase acts as a catalyst in changing H_2O_2 to $H_2O + O_2$.

The ferrous (Fe^{++}) ion in the heme of hemoglobin and myoglobin binds O_2. Hemoglobin carries O_2 in the blood whereas myoglobin carries O_2 in muscle cells. Hemoglobin and myoglobin differ in the composition of their protein moieties. Also, hemoglobin consists of four polypeptide chains, each with its own heme group, while myoglobin contains one polypeptide chain.

Of the four hemoglobin chains, two of the polypeptide chains are called alpha and two are called beta. Thus, normal adult hemoglobin is mainly alpha2, beta2 (= hemoglobin A, or HbA) (Fig. 10.2). Embryos and fetuses contain other kinds of hemoglobin groups. HbF (in the fetus) is alpha 2, gamma 2; HbA2 (alpha2, delta2) is a small subgroup of the adult.

In the eastern end of Porphy's are the breakdown products of hemoglobin, including bilirubin and its derivatives, which are components of bile. When red cells get old and are destroyed, mainly in the spleen, the heme ring breaks and the Fe^{++} is released, forming **biliverdin,** an open-chain molecule with no iron. Biliverdin changes to **unconjugated ("indirect") bilirubin** by reduction of its central carbon. Faced with this shocking transformation, the indirect bilirubin molecule staggers into the circulation, but being insoluble in water, has to be carried to the liver by albumin. Then:

A. The liver cells take up the unconjugated bilirubin.
B. The liver cells transform it into **conjugated ("direct") bilirubin** by conjugating it with glucuronate. The resultant more polar molecule is now soluble in water.
C. The liver cells release the conjugated bilirubin into the biliary duct system, which carries it to the small intestine.
D. In the small intestine, the conjugated bilirubin is transformed into **urobilinogen,** some of which is excreted and some of which is reabsorbed and carried back to the liver and reexcreted into the bile. As urobilinogen is water soluble, it may be excreted in the urine, if blood levels are high enough. The same is true of conjugated bilirubin, but not of unconjugated bilirubin, which is relatively nonpolar.

The dark color of stool is partly a consequence of the conversion of urobilinogen (colorless) to **stercobilinogen** (colorless) to **stercobilin** (brown) in the intestines.

Conjugation of drugs with glucuronate is a very important reaction in the detoxification of certain drugs that are excreted through the liver.

Where Is Porphy's Hemeland?

The porphyrins are, among other things, precursors of heme. Heme production occurs virtually all over the body but is especially noteworthy in the bone marrow and liver, as heme is an important component of both hemoglobin and the cytochromes.

Bilirubin metabolism involves the liver, the intestines, and the bloodstream.

Vitamin B_{12} production occurs only in microorganisms, but the liver is a particularly important site of B_{12} storage.

Key Connections Of Porphy's Hemeland

A. Porphy's Hemeland originates with succinyl CoA from the **Krebs Cycle.**
B. The glucuronate used in conjugation comes from **Carbohydrateland.**

CHAPTER 9. THE INFIRMARY

An infirmary is necessary in Biochemistryland because of the park's many accidents. The items in stock include vitamins, hormones, minerals, drugs, and equipment for certain laboratory tests.

Vitamins

Vitamins are chemicals that are necessary in trace amounts for normal body function. They are not produced in sufficient amounts by the body and must come from external food sources. Their structures, like their diffuse locations in Biochemistryland, are generally diverse and unrelated, as shown in figures 9.1 and 9.2. Molecules that contain (or are) vitamins are indicated in green rectangles on the Biochemistryland map. There are actually many more loci on the map that contain vitamins, but which have not been included, to avoid cluttering the map. In particular, NADH (which contains niacin) has not been drawn in at many steps.

It is useful to divide vitamins into water soluble and fat-soluble groups:

Water-soluble vitamins (fig. 9.1)	**Fat-soluble vitamins** (Fig. 9.2)
B_1 (thiamine)	A (retinol)
B_2 (riboflavin	D (calciferol)
B_6 (includes pyridoxine, pyridoxal, and pyridoxamine)	E (tocopherol)
	K
B_{12} (cobalamin)	
C (=ascorbate, ascorbic acid)	
Folacin (=folate, folic acid)	
Biotin	
Niacin (=nicotinic acid)	
Pantothenate (=pantothenic acid)	

The Water Soluble Vitamins

Vitamin B_1 (thiamine): Mainly acts as Thpp (thiamine pyrophosphate), a prosthetic group for 2-ketoglutarate dehydrogenase, pyruvate dehydrogenase, and transketolase. Examples:

2-ketoglutarate ⟶ succinyl CoA (E-10)
Thpp + multienzyme complex

pyruvate ⟶ acetyl CoA (D-8)
Thpp + multienzyme complex

xylulose 5-P + ribose 5-P ⟶ glyceraldehyde 3-P +
Thpp + transketolase

sedoheptulose 7-P (D-3)

Vitamin B_2 (riboflavin): acts mainly as the coenzyme FAD (flavin adenine nucleotide) and FMN (flavin mononucleotide), which are used in many oxidation-reduction reactions in which hydrogen atoms are received or donated. Particularly noteworthy examples are their uses in the chain of electron tranport and in the cytochrome P450 hydroxylase systems.

FAD ⟶ $FADH_2$
FMN ⟶ $FMNH_2$

Vitamin B_6 (pyridoxine, pyridoxal, and pyridoxamine): is a coenzyme that prefers the world of amino acid metabolism. It is the prosthetic group for all **transaminases.** Amino acid transamination is a particularly important function. For instance:

pyruvate + glutamate ⟶ alanine + 2-ketoglutarate (E-7)
enzyme + pyridoxal phosphate (PyrP)

2-ketoglutarate + aspartate ⟶ oxaloacetate + glutamate (E-11)
enzyme + PyrP

histidine ⟶ histamine (F-11)
enzyme + Pyrp

homocysteine + serine ⟶ cystathionine (E-6)
enzyme + PyrP

Vitamin B_{12} (cobalamin): resembles the porphyrins in structure and is synthesized in Porphy's Hemeland, but only by microorganisms. It has a cobalt ion in its center rather than iron (as in heme) or magnesium (as in chlorophyll). It is part of the coenzyme used for:

a. rearrangements. Main example:

methylmalonyl CoA ⟶ succinyl CoA (D-11)
enzyme + B_{12}

b. a methylation reaction

homocysteine ⟶ methionine (E-6)
enzyme + B_{12}

Intrinsic factor is a glycoprotein produced in the stomach that facilitates transport of B_{12} from the gut to the general circulation. A deficiency of intrinsic factor, e.g. due to autoantibodies against this protein, may thus result in B_{12} deficiency.

Vitamin C: Perhaps the most famous role of vitamin C is its functioning in the conversion of collagen proline to collagen hydroxyproline. Vitamin C acts as an antioxidant that protects a variety of molecules from oxidation. As a reducing agent, it facilitates the absorption of iron in the intestines by reducing it to the more absorbable Fe^{++} form. Examples:

proline (and lysine) (in collagen) ⟶ hydroxyproline (and hydroxylysine) (F-12) enz. + vit C

tyrosine ⟶ homogentisic acid (C-8)
enz + vit C

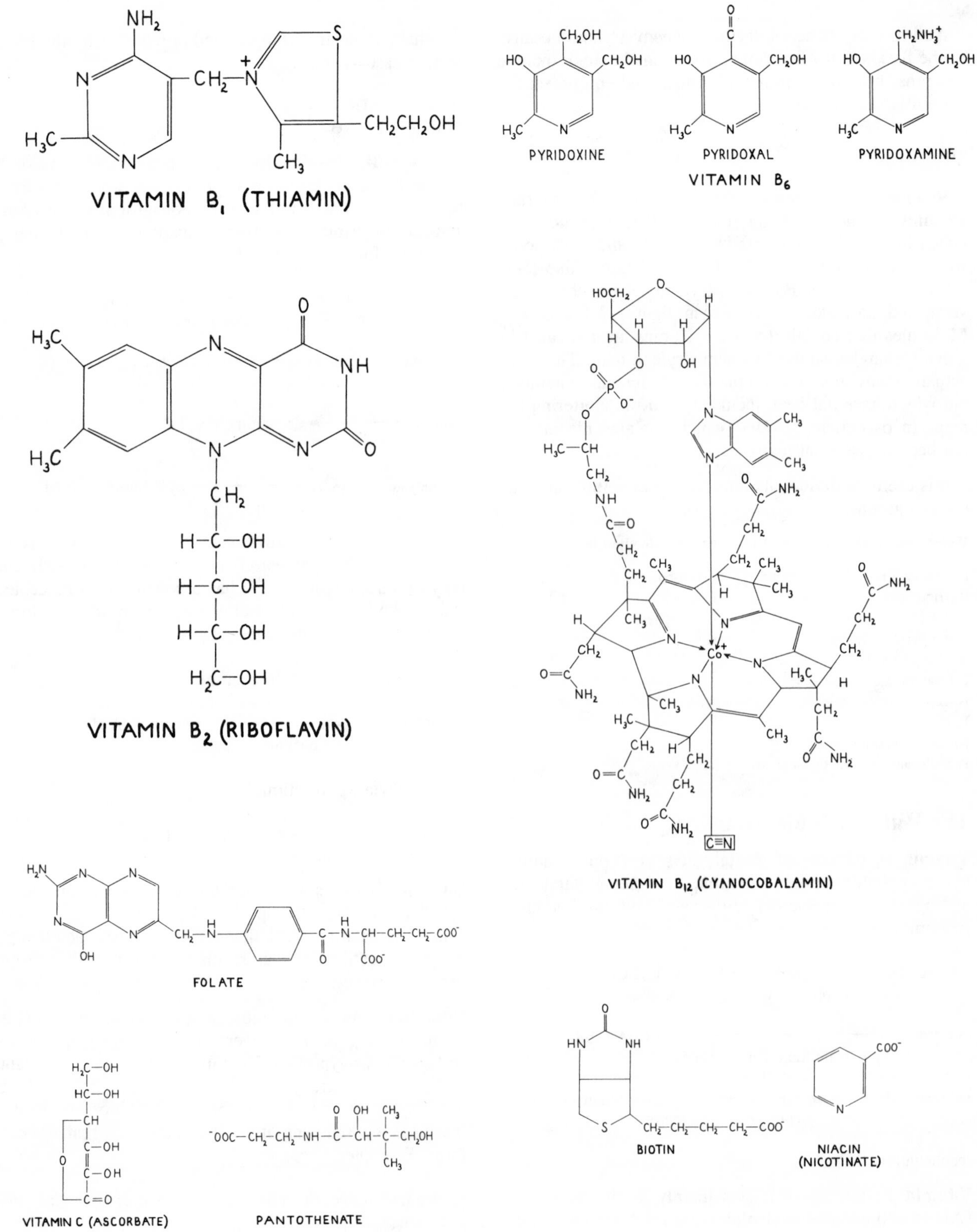

VITAMIN B_1 (THIAMIN)

PYRIDOXINE PYRIDOXAL PYRIDOXAMINE

VITAMIN B_6

VITAMIN B_2 (RIBOFLAVIN)

VITAMIN B_{12} (CYANOCOBALAMIN)

FOLATE

BIOTIN

NIACIN (NICOTINATE)

VITAMIN C (ASCORBATE)

PANTOTHENATE

Fig. 9.1. The water-soluble vitamins.

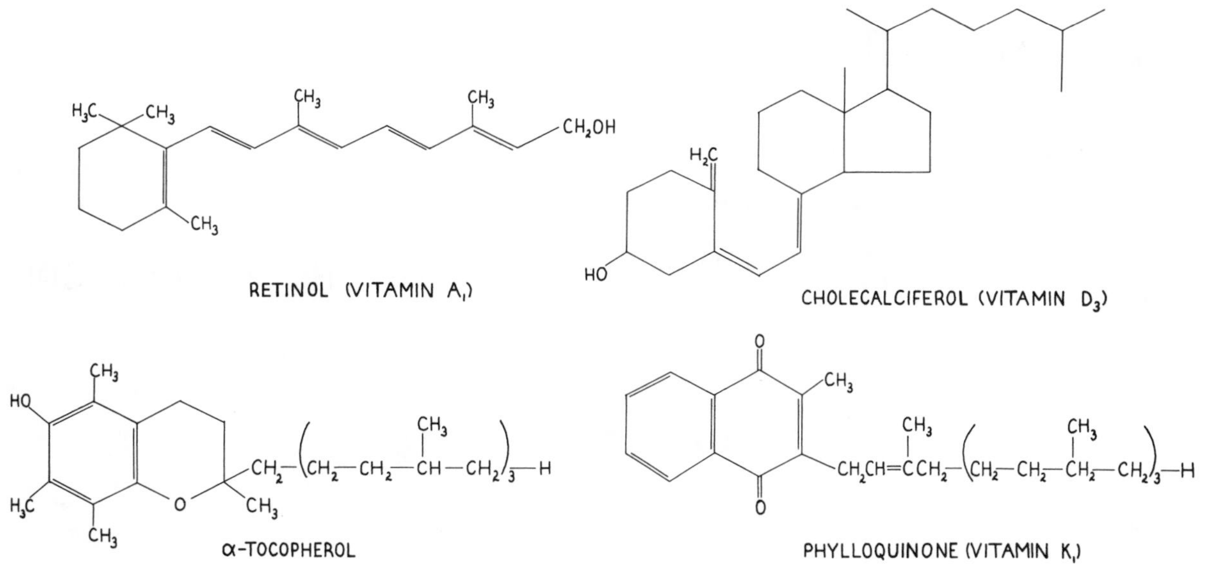

RETINOL (VITAMIN A_1)

CHOLECALCIFEROL (VITAMIN D_3)

α-TOCOPHEROL

PHYLLOQUINONE (VITAMIN K_1)

Fig. 9.2. The fat-soluble vitamins.

In all these reactions vitamin C works to keep the iron in the enzyme (hydroxylase) system reduced.

The adrenal gland, which produces epinephrine and norepinephrine, contains one of the highest concentrations of vitamin C of any body organ. It is believed that vitamin C may play a role in the synthesis of corticosteroids and epinephrine.

Folacin (=folate, =folic acid): is a precursor of THF (tetrahydrofolate), a 1-carbon donor in many Biochemistryland reactions. THF comes in a variety of activated (extra carbon-bearing) forms, each of which is capable of donating one carbon under certain conditions (fig. 9.3). Activated THFs contribute methyl groups in the formation of purine and pyrimidine rings (fig. 5.4), and hence THF is important for DNA synthesis. Also:

glycine (2 carbons) → serine (3 carbons)
(THF•C → THF)

homocysteine → methionine (E-6)
(THF•C → THF; enzyme + B_{12})

B_{12} and folacin are both important in rapidly dividing cells such as bone marrow cells. A deficiency of either B_{12} or folacin results in a megaloblastic anemia.

Biotin: is a coenzyme of **carboxylases.** It carries activated CO_2. For example:

pyruvate → oxaloacetate (D-8)
CO_2 + pyruvate carboxylase + biotin

acetyl CoA → malonyl CoA (G-8)
CO_2 + acetyl CoA carboxylase + biotin

propionyl CoA → methylmalonyl CoA (C-11)
CO_2 + propionyl CoA carboxylase + biotin

Niacin (=nicotinic acid). Just in case you're not convinced that vitamins are everywhere, niacin is part of the NADH and NADPH molecules and thus is vital to numerous reactions in Biochemistryland.

Pantothenic acid: is part of the CoA molecule (for instance, as acetyl CoA, malonyl CoA, and succinyl CoA), and part of the acyl carrier protein of fatty acid biosynthesis.

Can you find the vitamins that are hiding in the Main Powerhouse? They are there if you look hard. NADH derives partly from niacin; CoA contains pantothenic acid. Thpp, which is important in the change of pyruvate to acetyl CoA, contains thiamin. Riboflavin is found in FAD and FMN. ATP does not contain a vitamin.

The Fat Soluble Vitamins

Vitamin A (retinol) is present in many cells of the body. In the retina it is part of the rod photoreceptor pigment **rhodopsin** (=cis-retinal + a protein called **opsin**) and cone photoreceptor pigment **iodopsin** (=cis-retinal + a different opsin protein). When light strikes the cis-retinal part of the rhodopsin or iodopsin molecule, the "cis" form changes to trans-retinal, a conformational change in the molecule that amazingly initiates the chain of neuronal impulses (G-11). Trans-retinal subsequently

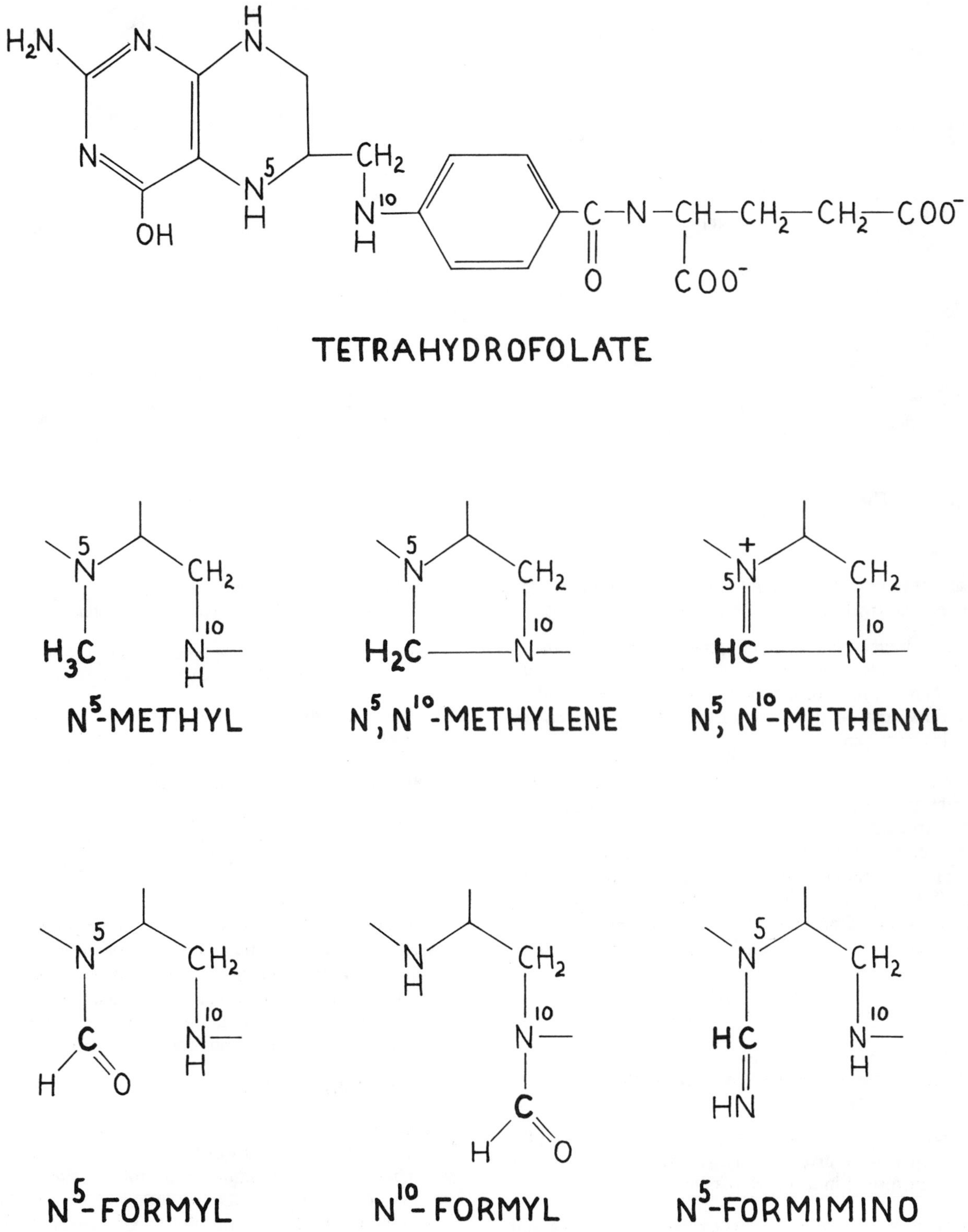

TETRAHYDROFOLATE

N^5-METHYL

N^5,N^{10}-METHYLENE

N^5,N^{10}-METHENYL

N^5-FORMYL

N^{10}-FORMYL

N^5-FORMIMINO

Fig. 9.3. Six "activated" (extra-carbon-carrying) forms of tetrahydrofolate, which acts as a 1-carbon donor.

back to the "cis-" form. Amongst its other functions, vitamin A helps to maintain skin and mucosal surfaces and facilitates the modeling of growing bone. It is believed that vitamin A, and its metabolite retinoic acid, may act like an antiproliferative hormone, entering cell nuclei and affecting DNA, resulting in alterations in gene expression.

Vitamin D, apart from its availability in the diet, is produced in the body, in response to sunlight striking the skin, and really acts like a hormone. It helps to raise the blood levels of calcium and phosphorus in several ways. It stimulates calcium and phosphorus absorption in the gastrointestinal tract; promotes transfer of calcium and phosphorus from bone to blood; promotes calcium retention by the kidney. As a hormone it acts by entering the cell nucleus and interacting with DNA to regulate protein synthesis.

Vitamin E appears to act as a antioxidant. It contributes electrons to lipids that are in the free radical form, stabilizing them and protecting them from oxidation.

Vitamin K is produced by plants (vit. K_1) or by bacteria (vit. K_2) including bacteria in the human intestine. It is necessary for the production of prothrombin and other clotting factors in the liver. It acts as a coenzyme in the carboxylation of several clotting factor proteins.

The effects of the various vitamin deficiencies are discussed in the Infirmary section of the Clinical Review.

Hormones

Hormones, like vitamins, are a grouping of very diverse structure. Unlike vitamins, they are produced in the body and generally travel, typically by the blood stream, to affect distant target organs that contain specific receptors on which the particular hormone acts. A receptor may lie within the cell membrane or more deeply within the cytoplasm or nucleus. Hormones are not coenzymes or enzymes but do modify the actions of enzymes.

The distinction between vitamins and hormones is not always that clear. What is a vitamin to us may be a hormone in the plant that produces it. What we call hormones generally appear to have a more profound effect on the body than do vitamins when their blood levels change slightly. Hence, hormones, which are produced in the body, cannot be bought over the counter, whereas vitamins, which are produced outside the body, can. Vitamin D, while called a vitamin, is really a hormone in that it is produced in significant quantities in our skin on exposure to sunlight.

Some hormones are proteins or polypeptides (oxytocin, TSH, insulin). Others, while neither proteins nor polypeptides, are derivatives of amino acids (thyroxine, epinephrine). Others (steroids) are produced in Lipidland as derivatives of cholesterol.

Hormones may act by affecting the rate of synthesis of proteins, such as enzymes, the activity of enzymes, or the permeability of cell membranes.

In general, the protein and polypeptide hormones, including epinephrine, interact with receptors on the cell surface. Specific cell receptors are important to insure that the hormone acts only on specifically designated cells. Once having reacted with a specific cell surface receptor, this stimulates a second messenger within the cell to regulate enzymic activity. For most protein and polypeptide hormones the second messenger is cyclic AMP. A good example is the cascade that results in phosphorylase activation in glycogen breakdown; glucagon uses cyclic AMP as a second messenger to activate glycogen breakdown in the liver. The second messenger in hormone actions is not always cyclic AMP, however. Other second messengers include cyclic GMP, inositol triphosphate, diglyceride, and Ca^{++}. Insulin is a polypeptide that has a cell-surface receptor but its action beyond that point does not appear to be mediated by cyclic AMP.

The steroid hormones act on cytoplasmic receptors, and the resulting complex moves to the nucleus to influence DNA transcription of proteins. Thyroid hormone directly enters the nucleus where it alters DNA, causing production of certain enzymes. A particular chemical may have more than one receptor type (e.g., alpha and beta receptors). Epinephrine, for instance, acts on alpha receptors (in skin and digestive tract arteries) to produce vasoconstriction, and on beta receptors (in skeletal and cardiac arteries and heart muscle) to produce vasodilation and increased cardiac ventricular contraction and heart rate.

Specific Hormones

GROWTH HORMONE:
ORIGIN: Anterior pituitary gland (acidophil cells)
STRUCTURE: Polypeptide
FUNCTION: promotes growth, gluconeogenesis, lipolysis, protein synthesis.

PROLACTIN:
ORIGIN: Anterior pituitary gland (acidophil cells)
STRUCTURE: Protein
FUNCTION: Stimulates breast development and lactose synthesis in pregnancy, among other functions.

ADRENOCORTICOTROPHIC HORMONE (ACTH):
ORIGIN: Anterior pituitary gland (basophil cells)
STRUCTURE: Polypeptide
FUNCTION: Stimulates pregnenolone production and cortisol secretion in adrenal cortex.

LUTEINIZING HORMONE (LH) AND FOLLICLE STIMULATING HORMONE (FSH):
ORIGIN: Anterior pituitary gland (basophil cells)
STRUCTURE: Glycoproteins
FUNCTION: In males, LH stimulates testosterone synthesis in the testes, whereas FSH stimulates spermatogenesis. In females, LH and FSH are both necessary for

maturation of the ovarian follicle. Subsequently, LH induces development of the corpus luteum of the ovary.

THYROID STIMULATING HORMONE (TSH):
ORIGIN: Anterior pituitary (basophil cells)
STRUCTURE: Glycoprotein
FUNCTION: Controls production of thyroid hormone.

MELANOCYTE STIMULATING HORMONE:
ORIGIN: Middle lobe of pituitary
STRUCTURE: Peptide
FUNCTION: Promotes melanin pigmentation of the skin.

HYPOPHYSIOTROPIC HORMONES:
ORIGIN: Hypothalamus
STRUCTURE: peptides
FUNCTION: Stimulate release of hormones by the anterior pituitary.

ANTIDIURETIC HORMONE (ADH; Vasopressin):
ORIGIN: Hypothalamus (stored in posterior pituitary)
STRUCTURE: polypeptide
FUNCTION: Acts on the kidney to promote reabsorption of water back into the circulation.

OXYTOCIN:
ORIGIN: Hypothalamus (stored in posterior pituitary)
STRUCTURE: Polypeptide
FUNCTION: Stimulates release of milk in women who are lactating; stimulates uterine contraction.

ENDORPHINS and ENKEPHALINS:
ORIGIN: Brain. Endorphins are mainly produced in the hypothalamus. Enkephalins have a more widespread central nervous system origin.
STRUCTURE: Peptides
FUNCTION: They have opioid-like effects of analgesia and sedation.

MELATONIN:
ORIGIN: Pineal Gland
STRUCTURE: Derivative of tryptophan and serotonin
FUNCTION: May have a role in regulating adrenal and gonadal functioning.

THYROID HORMONE – T4 (thyroxine) and T3 (triiodothyronine):
ORIGIN: Thyroid gland
STRUCTURE: Amino acid derivative of tyrosine
FUNCTION: Increases body metabolic rate.

CALCITONIN (thyrocalcitonin):
ORIGIN: Thyroid
STRUCTURE: Peptide
FUNCTION: Decreases plasma calcium. Acts on bone by decreasing the activity of osteoclasts (cells that break down bone). Its net effect is opposite to that of parathyroid hormone.

PARATHYROID HORMONE (PTH):
ORIGIN: Parathyroid glands
STRUCTURE: Polypeptide
FUNCTION: Maintains the level of calcium in the blood, acting mainly on bone and kidney. In bone, PTH stimulates osteoclast cells to produce bone breakdown with release of calcium and phosphorus (PTH has a similar effect in this regard as vitamin D, but operates by a different mechanism; PTH acts through cyclic AMP). In the kidney, PTH increases calcium reabsorption and phosphate excretion (vitamin D, however, increases absorption of both calcium and phosphorus in the kidney).

EPINEPHRINE:
ORIGIN: Adrenal medulla
STRUCTURE: Derivative of tyrosine
FUNCTION: Epinephrine stimulates glycogen breakdown, lipid breakdown, and gluconeogenesis (the opposite of insulin). The adrenal medulla produces more epinephrine that norepinephrine. Norepinephrine, though, is the predominant neurotransmitter in postganglionic axons of the autonomic nervous system, where it mediates sympathetic, particularly catabolic (energy-expending; "flight-or-fight") responses.

GLUCOCORTICOIDS (cortisol being most important):
ORIGIN: Adrenal cortex
STRUCTURE: Steroid
FUNCTION: Promote gluconeogenesis and protein and fat breakdown; antiinflammatory; increase gastric acid production.
NOTE: Production is stimulated by ACTH from the pituitary gland.

MINERALOCORTICOIDS (aldosterone being most important):
ORIGIN: Adrenal cortex
STRUCTURE: steroid
FUNCTION: Stimulates kidney reabsorption of sodium back into the circulation, with loss of potassium.
NOTE: Aldosterone production requires the **renin-angiotension system:** Renin is an enzyme secreted by the juxtaglomerular cells of the kidney in response to decreased arterial blood pressure and blood flow. Renin stimulates conversion of the protein angiotensinogen to angiotensin I which then becomes angiotensin II. Angiotensin II stimulates the synthesis and release of aldosterone by the adrenal cortex.

INSULIN:
ORIGIN: Pancreas
STRUCTURE: Polypeptide
FUNCTION: Clears the blood of glucose; stimulates glycolysis and glycogen synthesis; promotes protein and fat synthesis; inhibits gluconeogenesis; facilitates uptake of glucose by cells.

GLUCAGON:
ORIGIN: Pancreas
STRUCTURE: Polypeptide
FUNCTION: Stimulates glycogen breakdown and gluconeogenesis in the liver.

ESTROGENS (Estradiol being the most important)
ORIGIN: Ovary
STRUCTURE: Steroid
FUNCTION: Necessary for development of secondary female characteristics. Needed for proliferation of the uterine endometrium during the early (preovulatory, or proliferative) phase of the menstrual cycle.

PROGESTERONE:
ORIGIN: Corpus luteum of the ovary
STRUCTURE: Steroid
FUNCTION: Prepares the endometrium to receive the fertilized egg during the postovulatory (progestational) phase of the menstrual cycle.

TESTOSTERONE:
ORIGIN: Testes
STRUCTURE: Steroid
FUNCTION: Development of male genitalia, male secondary sex characteristics, spermatogenesis, and libido. Androgens also promote skeletal and muscular development.

CHORIONIC GONADOTROPHIN (HCG):
ORIGIN: Chorion and placenta
STRUCTURE: Glycoprotein
FUNCTION: Prevents corpus luteum from involuting in pregnancy, allowing a rise of estrogen and progesterone.

GASTRIN:
ORIGIN: Gastric mucosa
STRUCTURE: Polypeptide
FUNCTION: Activates secretion of gastric acid, pepsin, and intrinsic factor (the vagus nerve stimulates gastrin release).

SECRETIN:
ORIGIN: Duodenal and jejunal mucosa
STRUCTURE: Peptide
FUNCTION: Inhibits gastric acid secretion; stimulates the pancreas to secrete water and bicarbonate; stimulates stomach pepsin secretion.

CHOLECYSTOKININ (pancreozymin):
ORIGIN: Duodenal and jejunal mucosal cells
STRUCTURE: Peptide
FUNCTION: Stimulates pancreatic secretion of enzymes; stimulates gallbladder contraction. Has other effects in ways similar to gastrin and secretin.

MINERALS, DRUGS, AND LABORATORY TESTS ARE DISCUSSED IN THE CLINICAL REVIEW IN THE **INFIRMARY** SECTION.

CHAPTER 10. CLINICAL REVIEW

The aim of the preceding chapters was to provide a basis for approaching the numerous clinical disorders that involve biochemistry. As an overview of the clinical problems in Biochemistryland, consider the general reaction:

$$\text{SUBSTRATE (S)} \xrightarrow{\text{Enzyme}} \text{PRODUCT (P)}$$

A deficiency of the enzyme will cause a decrease in P and an increase in S. The normal amounts of S and P may also be changed by altering the degree of ingestion (or production) of S or by altering the amount of excretion (or breakdown) of P. When amounts of S, enzyme, or P are altered, select clinical conditions may develop in Biochemistryland. Sometimes an excess of a substrate or product may be harmful (e.g., urate deposits in gout; NH_3 in defective liver function). At other times lack of the substrate or product may be damaging (e.g., decreased formation of hydroxyproline in scurvy). Laboratory diagnostic tests in Biochemistryland are commonly geared to detecting either changes in enzyme concentrations, or changes in concentration of substrate or product, or other chemicals that may be altered in the condition, such as hormones, vitamins, and electrolytes.

Treatment philosophy often consists in resupplying the missing chemical or reducing an excess of it. Drugs may sometimes be used to mimic the effects of a natural chemical that is needed, or, conversely, to compete with and inhibit the effects of the natural chemical which is acting in harmful ways. When the root cause of the problem is not known, the treatment may aim to affect a secondary manifestation of the condition (e.g., using antiinflammatory agents in arithritis). Sometimes the treatment is empirical, the mechanism of action of the therapeutic agent being unknown (e.g. the use of certain centrally-acting anesthetics). In the case of antibiotics and anti-tumor agents, the idea is to damage microorganisms and tumor cells to a greater degree than normal human cells. Treatment in part rests on a knowledge of how microorganisms and tumor cells differ in their metabolism from normal cells.

Certain diseases consist of puzzling assortments of symptoms affecting particular organs in peculiar ways that are not readily explained by the map. Why, for instance, does the child with Lesch-Nyhan Disease characteristically mutilate himself by biting his lips and fingers? Is it simply a hypoxanthine p-ribosyl transferase deficiency that is responsible for all this, or is there more? On the one hand it could be that a particular substrate-enzyme-product imbalance selectively affects particular organs because these chemicals have special functions in those organs. An isolated enzyme defect might disrupt quite a lot of different steps not only at the level of its substrate or product but at steps well behind or ahead of the substrate and product, steps that may have unique functions in particular cells or organs. It may also be that certain conditions involve mutations of a gene that controls other genes. There would then be alterations in the quantities of a number of proteins apart from the single enzyme identified on the map. The root of the disorder then may not lie solely at the particular step for which an enzyme deficit is known.

Despite the fact that much is known about specific organ localization of particular biochemical reactions, there is much refinement that is needed in this knowledge to properly understand why many of the syndromes strike in specific localizations in specific ways.

In any guided tour, one must first know where to begin. Biochemistryland, though, really doesn't have a beginning or end. Let us, however, arbitrarily begin in Carbohydrateland with glucose, that wonderous sun-formed molecule that connects in one way or another with all areas of Biochemistryland.

Carbohydrateland (and Glycolysis)

Energy can derive from glucose metabolism either through the HMP shunt, or through the Main Powerhouse via glycolysis and the Krebs cycle. Red blood cells, which lack mitochondria, do not have a Krebs cycle but obtain their energy through the HMP shunt (NADPH production) and glycolysis (ATP production). Enzymic defects in either **glycolysis** (all the way from glucose through pyruvate) or the **HMP shunt** may result in hemolytic anemias, among other things. Some of these are listed below, but the most common ones are **glucose 6-phosphate dehydrogenase (G6PD) deficiency** and **pyruvate kinase deficiency.**

#1. (D-2) **Glucose 6-phosphate dehydrogenase deficiency.** There is a defect at this step in the HMP shunt. The NADPH produced by the HMP shunt normally acts indirectly to reduce glutathione (F-12) in red blood cells. The reduced glutathione helps in preserving the red cell by reversing the oxidation of hemoglobin and other proteins. As glutathione is not reduced properly in G6PD deficiency, an attack of hemolytic anemia may be precipitated in these patients by ingestion of oxidizing drugs, like sulfonamides, antimalarials, such as primaquine, or uncooked fava beans. The abnormally oxidized hemoglobin may then precipitate as **Heinz bodies** in the red blood cells, and hemolysis (red cell destruction) occurs. G6PD deficiency protects against malaria as malarial organisms need NADPH and the HMP shunt for their growth. The diagnosis of G6PD deficiency can be made by enzyme assay for G6PD in red blood cells. G6PD deficiency is inherited as an x-linked recessive trait. Hence, it is more common in males. It is particularly common in blacks.

#2. (D-2) **Hexokinase deficiency** at this step. As glucose can not be phosphorylated, various glycolytic intermediates are deficient. This is associated with a hemolytic anemia.

#3. (D-2) **Glucose phosphate isomerase deficiency.** Associated with a hemolytic anemia.

#4. (D-3) **Phosphofructokinase deficiency (Type VII glycogen storage disease).** Phosphofructokinase is deficient at this step in muscle. Glucose cannot be used effectively and the patient experiences muscle cramps. Red blood cells normally have a form of this enzyme, too, and patients may have a hemolytic anemia and myoglobinuria. Muscle biopsy shows the enzyme defect. Also, there are increased muscle glycogen stores, as glucose is shunted toward glycogen in the absence of a properly functioning glycolytic pathway. There is also an increase in muscle fructose 6-phosphate and glucose 6-phosphate, but a decrease in fructose 1,6-diphosphate.

#5. (D-4) **Aldolase deficiency** at this step. Associated with a hemolytic anemia.

#6. (D-4) **Triose-phosphate isomerase deficiency.** Associated with a hemolytic anemia.

#7. (D-6) **Phosphoglycerate kinase deficiency.** Associated with a hemolytic anemia.

#8. (D-5) **Diphosphoglyceromutase deficiency.** This side step in the glycolytic chain is particularly important in red cell metabolism as 2,3-diphosphoglycerate (2,3-DPG) decreases the O_2 affinity of hemoglobin and stabilizes the deoxygenated form of hemoglobin. DPG mutase deficiency results in a **deficiency** of 2,3-DPG, with excessive affinity for O_2, and anemia. Conversely, pyruvate kinase deficiency (#10), leads to **excess** 2,3-DPG and low oxygen affinity, which may allow better delivery of oxygen to tissues, but is associated with an anemia.

#9. (D-7) **Enolase deficiency.** Associated with a hemolytic anemia.

#10. (D-7) **Pyruvate kinase deficiency.** The cell can not produce the ATP that normally is produced during this reaction. There is a hemolytic anemia. The diagnosis can be made by assay for pyruvate kinase in red blood cells. The inheritance is autosomal recessive.

The Glycogen Storage Diseases (GSD)

In the glycogen storage diseases, excess glycogen accumulates in the liver or muscle, or both (depending on whether it is liver, muscle, or a systemic enzyme that is defective). Glycogen accumulates either because it cannot be broken down properly or because excess glycogen is shunted into storage, as glucose 6-phosphate cannot be metabolized elsewhere. For instance, we have already seen the latter situation in muscle phosphofructokinase deficiency (type VII GSD—disease #4). In the latter, glucose 6-phosphate cannot be shunted through glycolysis and instead backs up into glycogen synthesis. Glucose 6-phosphatase deficiency (#11) is another example of shunting into glycogen synthesis.

#11. (D-2) **Glucose 6-phosphatase deficiency (Type I GSD; Von Gierke's Disease).** There is a deficiency at this step in the formation of glucose by the liver. Glucose 6-phosphate instead forms other things and the flow of reactions shifts to:

A. Carbohydrateland. Glycogen accumulates in the liver, which becomes enlarged.

B. The DNA Funhouse. There is excess uric acid due to shunting through the HMP (Penthouse Powerhouse) shunt to form purines. Uric acid stones may deposit in the urinary tract.

C. Main Powerhouse. There is buildup of pyruvate and lactic acid.

D. Lipidland. As the liver is not turning out glucose, alternate sources of energy must be increased. Triglycerides break down in fat cells to release fatty acids, with elevation of serum lipids. There may be fatty deposits (xanthomas) in various parts of the body. The excess pyruvate (pyruvic acid) and lactate (lactic acid) result in a serum acidosis.

The patient thus may have an enlarged liver, hypoglycemia, elevated uric acid, serum acidosis with elevated pyruvate and lactate, hyperlipidemia, and growth retardation.

Diagnosis may be confirmed by the absence of glucose 6-phosphatase on liver biopsy. The liver cells are full of glycogen. Also, the normal elevation in blood glucose from infusion of glucagon, epinephrine or galactose is absent.

#12. (C-1) **Deficiency of branching enzyme (Type IV GSD; Anderson Disease).** Here the liver produces an abnormal glycogen with very long chains. There is an enlarged liver, with liver damage (cirrhosis) and a poor prognosis. The diagnosis is made by assay for the deficient enzyme in liver biopsy, or by similar assay in leukocytes or cultured skin fibroblasts. The structure of the glycogen in liver biopsy is also abnormal.

#13. (D-1) **Muscle phosphorylase deficiency (Type V GSD; McArdle's Disease).** Liver phosphorylase is normal, but muscle phosphorylase is deficient. The patient cannot break down muscle glycogen and experiences muscle cramps and weakness with exercise. Muscle biopsy may confirm the enzyme defect. There is no significant rise in lactate in an ischemic exercise test.

#14. (D-1) **Liver phosphorylase deficiency (Type VI GSD; Hers' Disease).** There is glycogen accumulation in the liver and liver enlargement. There is growth retardation, as amino acids are shunted toward gluconeogenesis rather than growth.

#15. (D-1) **Liver phosphorylase kinase deficiency (Type VIII GSD).** The defect in this enzyme prevents activation of liver phosphorylase (see fig. 3.2). There is accumulation of glycogen in the liver and liver enlargement. Diagnosis is made on assay for the defective enzyme in liver biopsy or in leukocytes or erythrocytes.

#16. (D-1) **Deficiency of debranching enzyme (Type III GSD; Cori's Disease).** The hypoglycemia is not too bad as the outer chain of glycogen can at least be broken down. The 1–6 linkages cannot be broken, however, and short-chained glycogen accumulates. There may be liver enlargement, muscular weakness and cramps, and excess fat breakdown with hyperlipidemia. Diagnosis can be made on liver biopsy, which shows glycogen accumulation and the absence of debranching enzyme. The decreased ability to break down glycogen can be treated by frequent feeding and supplementation of the diet with protein, fructose, and galactose, which can be converted to glucose.

#17. (E-1) **Deficiency of alpha 1,4 glucosidase (lysosomal acid maltase) (Type II GSD; Pompe's Disease).** Maltase, apart from its location in small intestinal cells also has a form that is found in lysosomes throughout the body. Deficiency in lysosomal maltase results in glycogen accumulation, not only in the liver but in skeletal and cardiac muscle and other tissues. There is enlargement of the liver, heart, and tongue, with poor muscle tone and early death due to cardiac failure or respiratory infection. Diagnosis may be made by muscle biopsy, which shows glycogen accumulation and absence of maltase activity.

Fructose Metabolism

Deficiency in fructose metabolism, as in **essential fructosuria,** may be perfectly benign, for one is not interrupting the main line of the glycolytic chain. However, there is one condition, **hereditary fructose intolerance,** which does cause significant problems.

#18. (E-3) **Essential fructosuria (fructokinase deficiency).** This is a benign condition in which high fructose levels, while not toxic, may be detected in testing the urine for sugar. Both glucose and fructose are reducing substances, capable of reducing chemical reagents that are used for the urinary test for reducing sugars (e.g., like Clinitest tablets).

#19. (E-3) **Hereditary fructose intolerance (deficiency in fructose 1-phosphate aldolase).** Unlike essential fructosuria, this is a significant clinical disorder. It is believed that the fructose 1-phosphate accumulation in this disorder may be toxic to the liver and kidney. For one thing, accumulation of fructose 1-phosphate ties up the inorganic phosphorus needed to form ATP from ADP. Fructose 1-phosphate also inhibits liver phosphorylase. Liver cells then function poorly and there may be liver enlargement, jaundice, and hypoglycemia. There is also decreased renal function (protein in urine) and poor growth. Ingestion of fructose causes the hypoglycemia and vomiting. Treatment involves the removal of fructose from the diet. Even in normal individuals, it is unwise to administer large doses of fructose intravenously (the same is true, for similar reasons, for xylitol or sorbitol). When the latter are used as an alternative to glucose, the phosphate derivatives of these sugars, by holding on to inorganic phosphorus, tend to block the synthesis of ATP needed by the liver.

The Ice Cream Parlor

#20. (B-1) **Lactase deficiency.** Lactase is produced in the intestinal microvilli. Deficiency results in poor digestion of lactose, bloating, abdominal cramps, and diarrhea on ingestion of milk products. It may be hereditary, particularly in blacks and orientals, but may be acquired through a variety of intestinal diseases that affect the small bowel. Diagnosis may be made by small bowel biopsy or more simply by a lactose tolerance test, showing that blood glucose does not rise after oral administration of lactose. The condition may be treated by lactose restriction. Some degree of lactose intolerance occurs normally with aging.

#21. (B-1) **Galactokinase deficiency** at this step. Galactitol accumulates and there are galactitol cataracts (lens opacities). Diagnostically, red blood cells in the condition have decreased galactokinase activity.

#22. (B-1) **Classic galactosemia.** There is a **defect in the transferase (galactose 1-P-uridyl transferase)** that normally allows galactose 1-P to change to UDP-galactose. Thus, there is a backup in galactose metabolism with accumulation of galactose and galactose 1-phosphate in the liver, causing liver damage (cirrhosis). There may also be kidney failure and mental retardation. It is believed that galactose 1-phosphate has an especially toxic effect. There may be cataracts due to deposition of galactitol in the lenses. The diagnosis may be suspected in an infant with failure to thrive and sugar in the urine which does not turn out to be glucose. Newborn blood is commonly screened routinely for elevated galactose. Red blood cells have a decrease in the transferase activity. Treatment consists of reducing the intake of galactose in the diet.

#23. (D-3) **Essential pentosuria (deficiency of xylitol dehydrogenase).** A benign accumulation of xylulose develops, which may be confused with glucose when detected in the urine.

THE MAIN POWERHOUSE

A number of conditions affecting the Main Powerhouse have already been discussed in reviewing the anemias that may result when there are defects in glycolysis, through pyruvate (#1–#10). There are other conditions that may affect the Main Powerhouse at steps beyond pyruvate.

#24. (D-8) **Pyruvate dehydrogenase deficiency.** Children with this may develop lactic acidosis with the additional buildup of pyruvate and alanine. Symptoms include severe neurological disturbances. It may be treated by decreasing carbohydrate intake and increasing intake of ketogenic nutrients, i.e. nutrients that will bypass the pyruvate dehydrogenase step in forming acetyl CoA.

The Krebs Cycle

There appear to be a scarcity of known genetic diseases that affect the Krebs cycle directly. Perhaps this is because the Krebs cycle is so vital that genetic conditions affecting it are likely to be incompatible with life. There are certain rare myopathies that appear to be associated with defective mitochondrial function. The patient may have muscle weakness on arising and easy fatiguibility in certain muscle groups, and the mitochondria look abnormal or are excessive in number. It is not clear in many of these conditions where the difficulty lies, whether it is in the Krebs Cycle, oxidative phosphorylation, or some other aspect of mitochondrial function. Whatever the case, it is certainly possible in the adult to poison the Krebs cycle with various poisons:

#25. (D-10) **Cyanide poisoning** interferes with electron transport. Cyanide binds to the Fe^{+++} of **cytochrome oxidase** (cytochrome oxidase = cyt. a + cyt. a_3) and prevents oxygen from combining with cytochrome oxidase. Cyanide also combines with the iron of hemoglobin.

Carbon monoxide, like cyanide, poisons by combining with the heme of both cytochrome oxidase and hemoglobin.

#26. (D-10) **Antimycin A,** a fungal antibiotic, blocks oxidative phosphorylation at the step between cytochromes b and c_1.

#27. (E-9) **Fluoroacetate (rat poison)** acts by converting to fluorocitrate and then inhibiting aconitase at this step.

Arsenic, in its arsenate (AsO_4^{3-}) form prevents the sequential transformation of glyceraldehyde 3-P to 3-P glycerate (D-5). Arsenate resembles phosphate and can substitute for it in oxidative phosphorylation, and also attach to glyceraldehyde 3-P, in place of phosphate. The resultant unstable arsenate compound does change to the normal end result, 3-P-glycerate, but without the normal production of an ATP. This is poisonous.

#28. (C-10) **Fumarase deficiency.** There is a deficit in the transformation of fumarate to malate. The infant has developmental retardation, with abnormal neuromuscular function, lactic acidemia, and fumarate aciduria. The lactic acidosis may result from a backup of Krebs cycle function, all the way to lactate. Lactic acidosis may also be present in rare disorders of cytochrome oxidase activity. Diagnostically, there is a deficit in fumarase activity in assay of liver and skeletal muscle mitochondria.

The Saloon

#29. (F-8) **Excess alcohol intake** may result in hypoglycemia. One reason is that alcohol metabolism produces NADH when ethanol converts to acetaldehyde. An excess of NADH prevents lactate from being transformed to glucose (in gluconeogenesis) and may also lead to lactate accumulation (lactic acidosis). Hypoglycemia and aldehyde toxicity may in part be implicated in the **fetal alcohol syndrome,** which occurs in infants born to alcoholic mothers. There may be growth retardation and central nervous system defects.

Alcohol dehydrogenase catalyzes the conversion of ethanol to acetaldehyde. It also changes ingested methanol and ethylene glycol (antifreeze) to toxic products (formic acid and oxalate, respectively). Ethanol, in an emergency, can be used to treat such toxicity, as it competes for alcohol dehydrogenase. Acetaldehyde in itself may produce symptoms of excessive vasodilation, flushing, and tachycardia in sensitive individuals who have a particularly active form of alcohol dehydrogenase (common in Japanese and Chinese).

LIPIDLAND

Most of the important diseases of lipids actually involve lipoproteins and glycolipids and will be discussed below in Combo Circle. Diabetes, though, does have significant effects in Lipidland:

#30. (H-9) **Diabetic ketosis.** The mechanism of ketosis (ketone production) in diabetes is quite complex, but a simple explanation of part of the mechanism is as follows. Insulin normally responds to the fed state by clearing the blood of glucose. Insulin, in addition to promoting the entry of glucose into cells, stimulates glycogen storage, lipid storage (as triglycerides) and protein synthesis. Insulin stimulates the production of enzymes important to glycolysis and decreases the production of enzymes unique to gluconeogenesis. In diabetes, there is an absence of insulin effect; there is a tendency for gluconeogenesis to occur. A significant source of this glucose is oxaloacetate from the Krebs cycle (D-9). This depletes the Krebs cycle of oxaloacetate. Meanwhile, as glucose is not being utilized, triglyceride breakdown increases to provide an alternate source of acetyl CoA for the Krebs cycle (H-8). As oxaloacetate is depleted, some of this acetyl CoA cannot get into the Krebs cycle and accumulates to form ketones (H-9). As you will recall, excess acetyl CoA cannot avoid ketosis by forming glucose, as the step from pyruvate to acetyl CoA is essentially irreversible (D-8). Some of the fatty acids released from fat cell triglyceride depots travel to the liver for further processing. This overwhelms the liver cells with fatty acids, much of which cannot be processed quickly. A fatty liver may then develop, along with elevated blood lipids.

In starvation, there also may be an elevation of ketones for a similar reason: oxaloacetate is depleted as it tries to form glucose; triglycerides also break down, forming ketones. The patient does not develop a fatty liver as there are not enough calories being ingested to make lipid stores or to overwhelm the liver with fatty acids. In

Kwashiorkor, however, there may be a fatty liver. In this condition, the patient ingests sufficient calories but has inadequate protein intake. Proteins are necessary for the liver to form and release lipoproteins, vehicles for carrying lipids. Without protein, lipids may accumulate in the liver. The patient has an enlarged liver and distended abdomen, which is also partly due to **ascites** (fluid collection in the peritoneal cavity), that arises from hypoalbuminemia.

#31. (H-5) **Pancreatitis.** Pancreatic **lipase** is a lipolytic enzyme that releases fatty acids from glycerides. It, as well as proteolytic enzymes produced by the pancreas, are normally, as a cell-protective measure, kept in inactive (zymogen) form in the pancreatic cell. Pancreatitis can activate these enzymes with resultant pancreatic damage.

#32. (H-6) **Carnitine deficiency.** Fatty acid oxidation occurs in the mitochondria. Fatty acids are stored as triglycerides outside the mitochondria. Carnitine is a molecule that transports long chain fatty acids into the mitochondria for oxidation. In the very rare condition where carnitine is lacking (or the enzyme associate with the combination of fatty acid with carnitine is lacking), fats cannot be properly utilized for energy. The patient experiences muscle cramps on fasting or exercising. Note: This illustrates the importance of fatty acid oxidation in muscle function.

#33. (G-4) **Respiratory distress syndrome.** Lecithin (phosphatidylcholine) is an important membrane phosphoglyceride. In the lung, one of the lecithins (they come in a variety of forms) reduces surface tension in the pulmonary alveoli. Lecithin deficiency in the premature infant results in the respiratory distress syndrome, in which the alveoli are collapsed and there is difficulty with air exchange.

#34. (G-3) **Niemann-Pick Disease.** Sphingomyelin collects in the brain due to a deficiency in sphingomyelinase, which normally removes phosphorocholine from sphingomyelin. Among other things, there are mental retardation and early childhood death.

#35. (H-4) **Nerve gas.** It is not necessary to cripple the Main Powerhouse in order to poison someone. Nerve gas, as well as organophosphate insecticides, inhibit acetyl cholinesterase (AChE), an enzyme important in the degradation of acetylcholine to choline. This can cause paralysis through inhibiting AChE at the junction between peripheral nerves and muscle. Death occurs through respiratory paralysis. AChE inhibitors are also used in surgery as muscle relaxants. AChE, apart from its presence in nerve and blood cell membranes is also present in the serum where it is called **pseudocholinesterase.** Prior to administering AChE inhibitors in surgery, it is important to ascertain that the patient does not have low serum pseudocholinesterase levels. Otherwise, the administered AChE-inhibiting drug—which is normally counteracted to a degree by pseudocholinesterase, may have an excessively prolonged effect and cause respiratory paralysis. Pseudocholinesterase levels may be low in certain congenital conditions or may be lowered by certain medications. Phospholine iodide, for instance, is an AChE inhibitor that is sometimes administered as an eye drop for glaucoma. This medication should be discontinued several weeks before administering succinyl choline, or other AChE inhibitor, during surgery.

#36. (H-4) **Myasthenia gravis.** Patients with this condition produce auto-antibodies against acetylcholine receptors, resulting in poor communication between nerve and muscle. The patient may experience profound muscle weakness, especially after repetitive muscle contractions.

Channel #5

#37. (H-10) **HMG CoA reductase.** This is an important rate-limiting step in cholesterol synthesis. Drugs that act at this step to inhibit HMG CoA reductase can lower blood cholesterol. Normally, excess cholesterol inhibits HMG CoA reductase by negative feedback, providing a natural control mechanism for cholesterol synthesis. Hereditary differences result in differences in feedback effects. In certain people there is a marked increase in serum cholesterol on increasing cholesterol intake, whereas in others, there is little increase, as the feedback mechanism is functioning more actively.

#38. (H-5) **Refsum's disease.** There is inability to break down **phytanic acid** (a branched chain fatty acid). This leads to severe neurological symptoms.

Steroid Biosynthesis

Hormone defects at key point in the hormonal pathways may give rise to a variety of syndromes in which there are increased or decreased mineralocorticoid, glucocorticoid and/or androgen-estrogen function.

#39. (G-12) **Cholesterol desmolase.** A partial block at this control step reduces the levels of all the steroid hormones.

#40. (G-12) **Cytochrome P-450,** like cytochromes a,b, and c is part of an electron transport chain in mitochondria. However, rather than the electron transport associated with oxidative phosphorylation, it functions in **hydroxylation.** This is clearly important in the combination of O_2 with cholesterol and its derivatives, where hydroxyl groups are added at key steps (in the liver, and adrenal glands particularly). Cytochrome P-450 is also important in the hydroxylation, and, thus, detoxification of many drugs in the liver (e.g., phenobarbital). Cytochrome P-450 has also been implicated in inducing cancer by converting potential carcinogens into more active forms.

#41. (G-12) (H-13) **3-beta-hydroxysteroid dehydrogenase deficiency.** There is a decrease in mineralocorticoid and glucocorticoid production, and abnormal sexual development.

#42. (G-12) (G-13) **21-hydroxylase deficiency.** This is the most common hereditary enzyme defect in steroid biosynthesis. There is decreased glucocorticoid and mineralocorticoid production, as the enzyme is common to both pathways. This leads to increased ACTH production, which in turn causes adrenal hyperplasia and increased pregnenolone production. This results in increased androgen production and virilization. Therapy consists in administering glucocorticoids. This decreases ACTH production by negative feedback.

#43. (G-12) (G-13) **11-beta hydroxylase deficiency.** As in 21-hydroxylase deficiency, there is a decrease in mineralocorticoids and glucocorticoids, with virilization.

#44. (G-12) **18-dehydrogenase and 18-hydroxylase deficiency.** There is a deficiency of mineralocorticoids.

#45. (G-13) (G-13) **17-alpha hydroxylase deficiency.** There is a decrease in both glucocorticoid and androgen production. The mineralocorticoid path is open and this may cause hypertension through excess mineralocorticoids.

#46. (H-13) **17,20 lyase deficiency.** There is a decrease in androgen production.

Bile Acids

#47. (H-12) **Gallstones.** Most gallstones are composed mainly of cholesterol. Bile salts and phospholipids normally prevent the precipitation of cholesterol, but cholesterol stones may form when the cholesterol/bile salt-phospholipid ratio increases excessively. **Chenodeoxycholate** may be used as oral therapy for cholesterol gallstones. It not only provides an extra recirculating source of bile acids but inhibits the rate-limiting step in cholesterol biosynthesis.

THE AMINO ACID MIDWAY

There are numerous diseases that affect amino acid metabolism. Mental retardation is a common concomitant of many of these conditions, which, for the most part, are rare. The diseases may be subclassified according to the kinds of amino acids affected.

#48. (E-11) **Hyperammonemia.** Excess ammonia levels drive the reaction from 2-ketoglutarate toward glutamate, thereby depleting the Krebs cycle of 2-ketoglutarate and reducing ATP output. This, among other things, may be a factor in depressing brain function with hyperammonemia, as the brain is deprived of ATP.

Disorders of Aromatic Amino Acids

#49. (C-7) **Phenylketonuria (deficiency of phenylalanine hydroxylase).** Occasionally, the defect is not in the enzyme but in the ability to regenerate tetrahydrobiopterin, which is also necessary for the reaction. There is a buildup and excretion of phenylpyruvate in the urine, giving it a **mousy odor.** Mental retardation is a prominent feature. Diagnosis can be made by routine urine testing for phenylpyruvate or serum testing for elevated phenylalanine levels. The condition is treated with a diet low in phenylalanine. Sometimes, tetrahydrobiopterin deficiency may be treated by supplying biopterin.

#50. (C-9) **Alkaptonuria (defect in homogentisate oxidase)** at this step. There is a buildup of homogentisate, which spills out into the urine. The excess homogentisate polymerizes as it stands (especially in alkaline urine), causing a dark-colored urine. The condition is generally benign but may result in arthritis in later years. The polymer binds to collagen and may result in **ochronosis,** in which connective tissue acquires a darker color; the ears, for instance, may adopt a bluish coloration through transmitted light.

#51. (C-8) (C-10) **Tyrosinemia.** There are enzyme defects at steps in the metabolism of tyrosine. These result in the accumulation of tyrosine and its metabolites in the urine and serum. Liver and kidney dysfunction, and mental retardation are common. The condition may be treated by lowering tyrosine and phenylalanine intake. Vitamin C may be helpful as it is a cofactor for hydroxyphenylpyruvate hydroxylase at this step.

#52. (C-7) **Albinism (tyrosinase deficiency).** Tyrosinase deficiency at this step blocks steps necessary to produce melanin, and the individual therefore lacks melanin pigmentation in the skin, hair, iris, and retinal pigment epithelium. In **oculocutaneous albinism,** the individual is sensitive to bright light and sunburns easily. There is an increased incidence of skin cancer. Treatment consists in the avoidance of prolonged, direct exposure to sunlight. There is also a form called **ocular albinism** in which the eyes alone are involved. There is another form of albinism in which tyrosinase is present, but the mechanism is unclear.

#53. (C-6) **Parkinson's disease.** There is a deficiency of dopamine in the brain stem, particularly in the midbrain, where there is a marked loss of substantia nigra cells. This leads to clinical problems of slowness, stiffness, and tremor. It is unclear where the primary defect lies. It could be, for instance, that the root of the disorder is not a defect in the metabolic pathway for dopamine but some other defect that affects, in some other way, those cells in the brain stem that happen to contain dopamine. Parkinson's disease is treated with L-DOPA (the precursor of dopamine) rather than dopamine, as L-DOPA can cross the blood brain barrier, whereas dopamine cannot.

Pyridoxine (vitamin B_6, in the form of pyridoxal phosphate is a cofactor in the formation of dopamine from L-DOPA. It used to be thought that pyridoxine supplements would be helpful to treat Parkinson's disease. The opposite was found: vitamin B_6 apparently also enhances L-DOPA conversion to dopamine in other areas of the body. This means that less of the administered L-DOPA is available for entry into the brain. Therefore, B_6 treatment presently is contraindicated in Parkinson's disease.

Carbidopa is commonly added to L-DOPA in the treatment of Parkinson's disease. Carbidopa inhibits dopa decarboxylase (the enzyme active at this step in the formation of dopamine), but does not cross the blood brain barrier. Thus, the administered L-DOPA is free to act in the brain but is inhibited from acting in the periphery where it may have unwanted side effects, like nausea and vomiting.

Defects in amino acid transport include:

#54. (E-7) **Cystinuria.** This is a genetic disease affecting epithelial cell transport of cystine and certain other amino acids, resulting in cystine excess and cystine stones in the urine.

#55. (I-9) **Hartnup's disease.** There is a defect in the epithelial transport of neutral amino acids (e.g., tryptophan) leading to poor absorption and excess excretion of these amino acids. Clinical signs resemble those of niacin deficiency (tryptophan is a precursor of niacin), namely the 3 D's: Diarrhea, Dementia, Dermatitis. The condition responds to nicotinamide administration. **Fanconi's syndrome** is a more generalized defect in molecular transport, involving a multitude of amino acids, glucose, calcium, phosphate, proteins, and other molecules. There may be decreased growth and rickets.

#56. (I-11) **Carcinoid tumor.** Elevated 5-hydroxyindole acetate (5-HIAA) occurs in carcinoid tumors of the intestine, where argentaffin cells secrete excess 5-hydroxytryptamine (serotonin). The patient experiences flushing, diarrhea, hypotension, and bronchoconstriction. (It is possible that these symptoms may relate to other chemicals also produced by argentaffin cells).

#57. (I-8) **Excess xanthurenate** in the urine. Xanthurenate levels in the urine increase with vitamin B_6 deficiency because B_6 (as pyridoxal phosphate) is necessary for the further chemical transformation of 3-hydroxykynurenine. Pyridoxine deficiency may be detected by giving the patient a loading dose of tryptophan. If pyridoxine deficiency is present, there will be a detectable excess of xanthurinate in the urine. Oral contraceptives may increase urinary xanthurenate levels, possible because estrogens increase the level of tryptophan dioxygenase, which acts during the steps between tryptophan and 3-hydroxykynurenine. Isoniazid (an antibiotic used for tuberculosis) interacts with and inhibits pyridoxal phosphate. Thus patients taking isoniazid should take pyridoxine supplements.

Diseases of Branched Chain Amino Acids

#58. (C-11) (I-9) **Maple Syrup Urine Disease.** There is a block in the degradation of the branched chain amino acids. **Leucine, isoleucine, valine,** and their ketoic acids are elevated in the blood and urine. Assays for these chemicals can be done in the laboratory. The urine acquires a "maple syrup" aroma. Infants with the condition have a variety of neurologic problems, including mental retardation. The condition is treated by dietary restriction of the affected amino acids.

#59. (C-11) **Hypervalinemia.** There is a suspected defect in valine transaminase, which acts at one of the steps in valine metabolism.

#60. (I-9) **Isovaleric acidemia.** This is believed to be a defect in the step from isovaleryl CoA to beta-methyl crotonyl CoA, in the metabolism of leucine. Rather than a maple syrup odor, there is an odor of "sweaty feet". The patient has various neurologic disturbances and mental retardation. Isovaleric acid is elevated in the plasma. It is treated by restricting dietary intake of leucine.

#61. (I-9) **Methylcrotonic aciduria.** There is, among other things, excess excretion of beta-methylcrotonic acid, a substrate in the course of leucine breakdown. Possibly the defect may lie in the enzyme at this step. Administration of biotin, which normally functions at this step in leucine breakdown, may help improve the symptoms.

Diseases of Sulfur-Containing Amino Acids

#62. (E-6) **Homocysteinuria** (defect in **cystathionine synthase** at this step). This is the most common form of homocysteinuria. The enzyme defect leads to elevated levels of homocysteine, which can be detected in the urine. Serum methionine is also elevated. The clinical problems include dislocation of the lens, mental retardation, and various skeletal and neurologic problems. The mechanisms are unclear. Treatment may include administration of pyridoxine, decreasing dietary methionine and increasing cysteine.

#63. (E-5) **Hypermethioninuria (decrease in methionine adenosyl transferase)** at this step. The condition is relatively benign, whereas cystathionine synthase deficiency (see above) is not benign.

#64. (E-6) **Variant of homocysteinuria.** In this variant, homocysteine cannot be converted to methionine. The ineffectiveness of the enzyme may be due to the inability of vitamin B_{12} to form the necessary enzyme cofactor, methylcobalamin. Alternatively, in other cases there may be a lack of the enzyme that forms 5-methyl THF (which is a methyl donor in the reaction).

#65. (E-6) **Cystathioninuria (deficiency of cystathioninase).** Cystathionine is elevated in the urine. Treatment with vitamin B_6 may reduce the urinary levels of cystathionine, but generally such treatment is unnecessary as the condition tends to be benign.

Other Amino Acid-Related Disorders

#66. (F-11) **Histidinemia (lack of histidase).** Histidine cannot be changed to urocanate. Blood and urine histidine levels are elevated, and urocanate, which normally is found in sweat, is absent. The patients may be mentally retarded. It may be diagnosed by checking blood and urine for histidine and its minor derivatives. On adding ferric chloride to the urine, "if the urine turns green there is histidine".

#67. (F-12) **Histamine** effect in allergy and shock. Histamine is widespread in the body organs. It is largely found in mast cells and basophils. It is released under a number of conditions, including the presence of certain drugs, tissue trauma, and antigen-antibody allergic interactions. It causes arteriolar relaxation, which may be helpful in increasing blood flow to injured tissues, but may result in hypotension and shock if excessive. Histamine excites smooth muscle, and this may cause bronchospasm. It stimulates sensory nerve endings, which may cause itching and pain. It also stimulates gastric secretion.

Chromolyn sodium, an adjunct in the treatment of asthma, acts to inhibit the release of histamine from mast cells. **Cimetidine** is a drug that resembles histamine in structure. It competes with histamine for receptors in the stomach, and is useful for reducing gastric acid secretion in the treatment of peptic ulcers.

#68. (F-13) **Formiminotransferase deficiency.** This results in increased urinary levels of FIGlu (formiminoglutamate). The patient may have significant neurological and other physical defects.

#69. (F-12) **Scurvy (deficiency of vitamin C).** Collagen is an important intercellular structural protein that contains hydroxyproline and hydroxylysine, amino acids that are rare in other kinds of protein. The concentration of glycine is also very high in collagen. In scurvy there is defective collagen formation, as vitamin C is necessary for the hydroxylation of proline. Among other symptoms, the patient with scurvy bruises easily and has decaying gums.

Defects in collagen synthesis may occur at steps other than that of hydroxylation of proline. Steps in collagen synthesis include the following:

Intracellular steps:

A. Synthesis of the initial polypeptide in the fibroblast cell.

B. Hydroxylation (formation of hydroxyproline and hydroxylysine) and glycosylation of (addition of sugar to) the polypeptide (vitamin C is a cofactor that is necessary for the hydroxylation of proline).

C. Formation of a **procollagen triple helix,** each containing 3 peptide strands.

D. Release of the procollagen triple helix from the fibroblast.

Extracellular steps:

E. **TRImming** the ends of the procollagen **TRIple** helix to form a **TROpocollagen** triple helix.

F. Lining up of many tropocollagen triple helices to form a **collagen fiber.** The particular staggering of tropocollagen helices within the collagen fiber accounts for the periodicity of lines seen on viewing a collagen fiber through the electron microscope.

G. Cross-linking of tropocollagen bundles within the collagen fiber.

In **Ehlers-Danlos Syndrome,** which occurs in a variety of forms, there appear to be problems at later stages in the structuring of collagen. The trimming of the procollagen triple helix to form a tropocollagen triple helix may be defective. There may also be defects in collagen cross-linking. The patient has very elastic skin and is double-jointed.

In **lathyrism** in cattle, the defect is still later in the stages of collagen formation. There is a toxic interference with collagen cross-linking, on ingesting sweat peas.

In **osteogenesis imperfecta,** the child has "brittle bones", which fracture easily. There may also be blue sclerae, hearing defects, and dental abnormalities. There are a variety of forms of this condition, the bases of which are unclear but seem to involve defects in the structure of collagen.

Once formed, mature collagen can be digested by **collagenase.** Certain bacteria, e.g., Clostridium histolyticum, produce collagenase as a means of facilitating their spread through tissue planes. Collagenases are also important in the restructuring of tissues during growth and regeneration.

#70. (F-13) **Hyperhydroxyprolinemia.** It is postulated that there is a defect in the oxidation of hydroxyproline. The symptoms may include mental retardation in addition to elevated plasma hydroxyproline.

#71. (C-11) **Propionyl CoA carboxylase deficiency.** The serum shows elevated propionate. The patient may have mental retardation, ketoacidosis, protein intolerance and other defects.

#72. (D-11) **Defective metabolism of methylmalonyl CoA.** Infants with this may present with acidosis. There is an excess of methylmalonate in the urine, as methylmalonyl CoA cannot change to succinyl CoA. Sometimes this is due to a defect in vitamin B_{12} metabolism and sometimes to a defect in methylmalonyl CoA mutase, which is necessary at the same step as Vitamin B_{12}.

The Urea Rest Room

In general, defects of the urea cycle give rise to elevated levels of ammonia, as the ammonia cannot effectively enter the urea cycle for elimination. There results a variety of neurologic disturbances, including mental retardation. Treatment may consist of lowering protein intake. Treatment with benzoic acid may facilitate ammonia elimination, through alternate pathways that involve hippuric acid production.

#73. (B-9) **Carbamoyl phosphate synthetase deficiency.** Ammonia cannot effectively enter the urea cycle and there is hyperammonemia.

#74. (B-9) **Ornithine transcarbamoylase deficiency.**, another cause of hyperammonemia.

#75. (A-9) **Argininosuccinate synthetase deficiency.** There are elevated citrulline levels in the blood, urine, and cerebrospinal fluid, and there may be hyperammonemia.

#76. (A-9) **Argininosuccinic aciduria.** There is decreased argininosuccinase activity. Argininosuccinate is elevated in blood and urine, whereas arginine is lower. For some reason, the hair is very dry and brittle.

#77. (A-10) **Hyperargininemia.** There is defective arginase activity. Blood arginine levels are increased.

COMBO CIRCLE

Clinical disorders may occur in the Lipoprotein, Glycolipid, or Glycoprotein sections of Combo Circle.

#78. (F-2) **Lipoprotein disorders.** The most common causes of hyperlipidemia are diseases of lipid metabolism and excess alcohol intake (alcohol is converted to acetyl CoA, which can convert to lipids).

There is considerable interest in the variety of disorders involving lipid metabolism, particularly in view of the correlation between cardiovascular disease and increased serum cholesterol and triglycerides. Hyperlipidemia may present in a variety of ways, such as plaques **(atheromata)** in blood vessel walls in atherosclerosis, fatty skin lesions in the eyelids **(xanthelasmae),** and lipid deposits in tendons **(xanthomas)** and the cornea **(corneal arcus).** Elevated cholesterol appears to correlate better than elevated triglycerides with atheromata, corneal arcus, xanthelasmae and certain xanthomata. Other kinds of xanthomata, as well as turbid plasma, coincide better with elevated triglycerides.

Pancreatitis may result from elevated triglycerides, as pancreatic lipase can act on elevated blood triglycerides to form breakdown products that may injure pancreatic cells.

Cholesterol is found in all lipoprotein particles but is relatively concentrated in LDL and HDL, whereas triglycerides are relatively concentrated in chylomicrons and VLDL. Heart disease appears to correlate with increased LDL and decreased HDL. Elevated HDL generally is a favorable finding, as HDL shunts excess cholesterol back to the liver (via LDL remnant particles) where it can be excreted. Chylomicrons, VLDL, and LDL, on the other hand, carry triglycerides and cholesterol to the periphery (fig. 6.4). The more significant clinical conditions involve elevation of cholesterol, triglycerides, VLDL, LDL, and/or chylomicrons, rather than HDL.

Treatment of lipid disorders includes:

a. decreasing dietary intake of cholesterol and saturated fats. Unsaturated fats appear to decrease VLDL and raise HDL. HDL levels are also increased by prostaglandin precursors in certain fish oils, a topic of significant current interest.

b. use of cholestyramine or colestipol, which bind bile salts and prevent the bile salts (which are cholesterol derivatives) from being reabsorbed. In extreme cases, one can remove a portion of the small intestine to reduce fat absorption.

c. nicotinic acid, which decreases VLDL production.

d. inhibitors of HMG CoA reductase to decrease the production of cholesterol (see #37).

e. clofibrate, which activates lipoprotein lipase to help clear VLDL and chylomicrons. It also appears to reduce cholesterol and triglyceride biosynthesis.

f. Ingesting medium chain fatty acids. In lipoprotein lipase deficiency there is a deficit in the breakdown of triglycerides. However, **medium** chain fatty acids normally are absorbed right into the portal system and travel attached to albumin, without being converted into triglycerides. Medium chain fatty acids therefore can be administered without their requiring lipoprotein lipase for their processing.

The functioning of the lipoproteins depends in part on their specific apoproteins. A defect in an apoprotein may lead to a decrease in the production of a particular lipoprotein (e.g., familial hypolipoproteinemia). Alternatively, an apoprotein defect may cause a rise in lipoproteins: if the apoprotein is necessary for allowing the lipoprotein to recognize its target site, the defect may prevent the lipoprotein from unloading its lipids and there will be a lipoprotein buildup in the blood.

In order for lipoproteins to release their cholesterol and triglycerides, the cholesterol receptors on the target cells and lipoprotein lipase in the fat and muscle capillary walls must be intact. If there is a deficiency in these, the corresponding lipoprotein will be elevated. In familial lipoprotein lipase deficiency, there may be elevation of chylomicrons and VLDL. LDL receptors are defective in familial hypercholesterolemia, and there may be elevation of LDL.

Lipoprotein disorders may also result from a variety of intestinal disorders that impede lipid absorption (such as certain inflammatory bowel diseases). Lipid disorders may also occur when there is triglyceride breakdown and shifting of triglyceride stores, as in the lipidemia that may accompany excess lipid breakdown in diabetes.

Some of the lipoprotein disorders, apart from dietary-induced, that illustrate the above principles are (see also fig. 6.4):

A. **Familial lipoprotein lipase deficiency** (Type I lipoprotein pattern on electrophoresis). Serum triglycerides become elevated with particular elevation of chylomicrons. There are xanthomas, rather than atherosclerosis. Pancreatitis may result from the action of pancreatic lipase on these elevated chylomicrons, with resultant excess triglyceride breakdown in the pancreas, pancreatic injury, and release of more pancreatic lipase. (Note that the body contains different kinds of lipases. There is a **pancreatic lipase,** which is a digestive enzyme; a **lipoprotein lipase,** which is an extracellular enzyme that breaks down plasma triglycerides, thereby enabling fatty acids to enter cells; and an **intracellular lipase** that breaks down stored triglycerides).

B. **Familial hypercholesterolemia** (Type IIa or IIb pattern). Elevation of LDL (sometimes VLDL, too). This involves a defect in the cell LDL receptor site.

C. **Familial dysbetalipoproteinemia** (Type III pattern). A defect in the remnant particle apoprotein, which results in the loss of ability of remnants to bind to liver cells. There are xanthomas and marked premature atherosclerosis. Remnant particles accumulate in the plasma, seen as a broad beta band on electrophoresis. There is elevation of both plasma cholesterol and triglycerides.

D. **Familial hypertriglyceridemia** (Type IV or V pattern). There is elevated VLDL, but the mechanism is unclear (possibly a defect in VLDL catabolism). There may be xanthomas, pancreatitis and premature atherosclerosis.

E. **Multiple lipoprotein-type hyperlipidemia** (Type IIa, IIb or IV pattern). Elevated LDL, possibly due to excess production of VLDL. There is premature atherosclerosis.

Conditions F, G, and H below are associated with **hypolipoproteinemia** and a different assortment of clinical problems.

F. **Abetalipoproteinemia.** There is absence of apoproteins. Chylomicrons are not found and there is fat malabsorption. There are various neurologic abnormalities and distorted, "thorny" erythrocytes **(acanthocytosis).**

G. **Familial hypobetalipoproteinemia.** There is a decrease but not a total absence of apoprotein B (apoB) and its lipoprotein. It is relatively benign clinically.

H. **Tangier disease.** There is an absence of HDL due to a defect in its synthesis or catabolism. Serum cholesterol levels are low. Cholesterol cannot be transported back to the liver and, instead, accumulates in phagocytic cells. The tonsils are orange (like a **Tang**erine) and enlarged. There are various neurologic problems and corneal opacities (but no premature atherosclerosis).

I. **LCAT deficiency.** Cholesterol associated with HDL cannot be esterified. There is a buildup of unesterified cholesterol, with corneal opacities, renal insufficiency, hemolytic anemia, and premature atherosclerosis. The diagnosis may be made on enzyme assay for plasma LCAT.

Glycolipid Theatre

A common denominator finding in glycolipid disorders is the **inability, in lysosomes, to break down particular glycolipids,** resulting in glycolipid accumulation. Clinical findings commonly include mental retardation.

#79. (G-1) (G-2) **Tay-Sachs disease.** There is a deficiency in hexosaminidase A, which normally breaks down gangliosides. This results in excess accumulation of **ganglioside** in the brain of the infant. The child appears normal at birth but experiences progressive neurologic deficits, leading to death within a few years.

#80. (G-1) (G-2) **Gaucher's disease** (deficiency of beta-glucosyl ceramidase, which normally breaks down glycosyl cerebroside to ceramide). This results in the accumulation of **glucosylcerebroside** in the brain, liver and spleen. There is hepatosplenomegaly in addition to neurologic deficits. The diagnosis may be made by examining the bone marrow for "Gaucher's cells", which contain excess cerebroside. Leukocytes may also be assayed for the missing enzyme and altered histological appearance. The condition may occur in infancy or later life.

In **Krabbe's disease (globoid leukodystrophy)** there is accumulation of **galactocerebroside** in the white matter of the child's nervous system, as lysosomes lack the enzyme galactocerebrosidase. There is poor myelinization, optic atrophy, mental retardation and, typically, death within 1–2 years of the condition's onset in infancy.

Fabry's disease is an X-linked disorder of globoside breakdown, usually in young adults. There is a defect in lysosomal alpha-galactosidase A. The patient has a painful neuropathy and, among other things, progressive renal failure with accumulation of globoside in the kidneys.

Glycoprotein Theatre

#81. (F-1) The **mucopolysaccharidoses** are proteoglycan disorders that generally result from a hereditary lysosomal defect in enzymes that normally degrade mucopolysaccharides (in most cases heparan sulfate and dermatan sulfate). This leads to the accumulation of different mucopolysaccharides, which may be associated with a variety of different findings, commonly including mental retardation and various skeletal abnormalities. These diseases include **Hunter disease, Hurler and Scheie disease, "I-cell disease", Maroteaux-Laury disease, Morquio syndrome, Mucolipidoses VII disease, multiple sulfatase deficiency,** and **Sanfilippo A and B diseases,** which will not be elaborated on further here. Often these conditions can be detected in advance on amniocentesis.

In **pulmonary alveolar proteinosis,** the alveoli of the lung fill with an excess of a glycoprotein. The reaction may occur in response to dust or other irritants.

THE DNA FUNHOUSE

There are a number of disorders that present at the level of purine and pyrimidine metabolism, the most common one being **gout,** which results in the abnormal deposition of urate crystals. At the level of the nucleic acids, the most important pathology is that of the enormous numbers of gene mutations that may affect the kinds and amounts of proteins that are produced. These proteins include enzymes, and it is clear that most of the clinical diseases in Biochemistryland, since they involve enzyme abnormalities, involve gene mutations. Mutations may alter an enzyme's function by resulting in a new enzymne with different functional capacities. Alternatively, if the mutation occurs in a gene that controls other genes, the enzyme in question may be normal in structure, but produced in abnormal amounts.

Disorders of Purine Metabolism

#82. (B-5) **Gout.** In gout, there commonly is an elevation of uric acid in the blood and sometimes in the urine. Urate crystals precipitate, particularly in joints (tophi, especially in the big toe) and urine (urate stones). Hyperuricemia (elevated blood uric acid) may result from excess production of uric acid and/or decreased excretion. Causes of excess production include metabolic defects that shunt the pathways toward uric acid production, e.g., Lesch-Nyhan syndrome—#84 (B-5), PRPP synthetase overactivity—#83 (C-4), and glucose 6-phosphatase deficiency—# 11 (D-2). Various malignancies may also result in elevation of uric acid through cell breakdown and conversion of excess nucleic acids to uric acid. Renal disease may block the excretion of uric acid and also result in hyperuricemia. Alcohol may precipitate an attack of gout by decreasing uric acid excretion and increasing production.

#83. (C-4) **Overactivity of either PRPP synthetase (which acts between ribose 5-P and PRPP) or amidotransferase** (which acts between PRPP and 5-P-ribosylamine). A modification of these enzymes may lead to their overactivity, with consequent hyperuricemia.

#84. (B-5) (B-5) **Lesch-Nyhan syndrome (deficiency of hypoxanthine P-ribosyl transferase).** This is an X-linked defect in the **salvage pathway** that normally changes hypoxanthine and guanine to IMP and GMP. In the absence of this salvage, there are urate and PRPP accumulation (PRPP normally combines with hypoxanthine and guanine to form IMP and GMP) and, subsequently, gout. There are also puzzling neurologic signs—mental retardation and self-mutilating behavior such as biting the lips and fingers.

#85. (B-5) **Xanthine oxidase deficiency.** The inability to change xanthine to urate results in xanthinuria with decreased blood and urine urate levels. Sometimes urinary xanthine stones may form. The absent enzyme may be confirmed on liver biopsy.

#86. (B-4) **APRT (adenine p-ribosyl transferase) deficiency.** The absence of APRT at this step between adenine and AMP results in shunting of adenine toward another product—2,8-dioxyadenine, which may precipitate as urinary stones.

#87. (B-4) **Adenosine deaminase deficiency.** This is associated with low numbers of lymphocytes and severe combined immunodeficiency disease. There appears to be an excess production of ATP and dATP, which leads to an imbalance that upsets the proliferation of immune cells.

#88. (B-5) **Purine nucleoside phosphorylase deficiency.** Associated with immune system deficiency and decreased numbers of T lymphocytes.

Disorders of Pyrimidine Metabolism

#89. (B-8) **Orotic aciduria.** Blocks in steps leading to pyrimidine synthesis may result in deficient production of pyrimidine nucleotides. There may be anemia and immune deficiency (from decreased red and white cell production) and excess orotic acid which may precipitate in the urine. Treatment with uridine may not only help supply the missing pyrimidine, but can decrease the level of orotate by uridine's negative feedback on steps that lead to orotate production.

#90. (B-7) (B-7) **Pyrimidine 5′-nucleotidase deficiency.** RNA, in this condition, can not be completely degraded in maturing red blood cells. The nucleotides of uridine and cytidine accumulate. This is associated with a hemolytic anemia. In **lead poisoning,** lead inhibits 5′-nucleotidase and also may result in an anemia. In both lead poisoning and pyrimidine 5′-nucleotidase deficiency, the red cells have the microscopic appearance of "basophilic stippling", which may be due to incompletely degraded RNA. Lead also inhibits the incorporation of iron in heme synthesis (#99).

#91. (A-6) In **xeroderma pigmentosum** there is a defect in the normal DNA repair process. Patients with the disease are susceptible at an early age to developing skin cancer following sun exposure, which can cause abnormal pyrimidine dimers in the DNA molecule. The condition may involve a deficiency in an endonuclease.) **Ataxia telangiectasia** and **Fanconi's syndrome** are conditions that are also believed to be related to endonuclease deficiency.

In the former, there are abnormal proliferation of blood vessels in addition to cerebellar dysfunction. In Fanconi's syndrome, there is a defect in the renal tubule transport of many kinds of molecules, including amino acids, monosaccharides, proteins, uric acid and electrolytes.

#92. (A-6) **Proteins disorders** form a vast category that includes not only the numerous **enzyme defects** in Biochemistryland but many other defects of proteins that are not enzymes. These are far too numerous to describe in detail, but include: diseases of transport proteins (e.g., the hemoglobinopathies, hypoalbuminemia): disorders of the immune system (abnormal antibodies in autoimmune and other immune disorders—fig. 10.1); blood clotting disturbances (e.g., the hemophilias, excess thrombosis); abnormalities of cell membranes (e.g., defective cell surface receptors); and abnormalities of those hormones that are proteins. Many of these disorders are primary defects that stem from gene mutation. Others may be acquired.

PORPHY'S HEMELAND

Various diseases may affect porphyrin metabolism on the way to forming heme. These result either in the accumulation of porphyrin derivatives in red blood cells **(erythropoetic porphyrias)** or in the liver **(hepatic porphyrias),** as the liver is also a site of porphyrin production (heme is important not only in hemoglobin but in the cytochromes). Curiously, anemia often is not a strikingly prominent characteristic of these diseases, but other things are common—photosensitivity with skin scarring (porphyrins absorb light), abdominal pain, and various

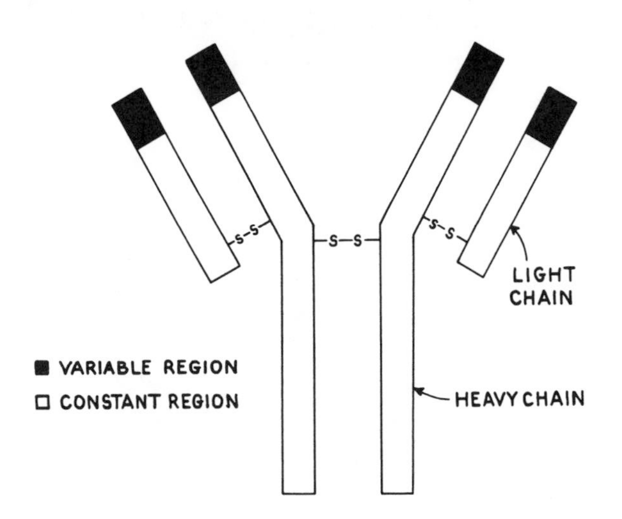

Fig. 10.1. An immunoglobulin (antibody) molecule. Each molecule contains two heavy and two light protein chains, each of which contains a **variable** portion, which is responsible for the subtleties of immunologic specificity. The type of heavy chain determines the main immunologic class (e.g., IgG, IgA, IgM, IgD, IgE).

neurologic problems. Examination of the red blood cells, urine and/or feces for various porphyrin products is useful in making the specific diagnosis. The urine is often pink or red, due to porphyrin excretion. Enzyme defects may occur at a number of different steps in porphyrin synthesis, but fine differentiating details of the conditions will not be discussed:

#93. (D-11) **Vitamin B_6 deficiency.** By reducing pyridoxal phosphate availability at the entrance to Porphy's Hemeland, an anemia may result through decreased ability to form heme.

#94. (D-12) **Uroporphyrinogen I synthase deficiency** (acute intermittent porphyria).

#95. (C-12) **Uropophyrinogen III cosynthase deficiency** (congenital erythropoetic porphyria).

#96. (D-13) **Uropophyrinogen decarboxylase deficiency** (porphyria cutanea tarda).

#97. (D-13) **Coproporphyrinogen oxidase deficiency** (variegate porphyria).

#98. (D-13) **Protoporphyrinogen oxidase deficiency.**

#99. (D-12) **Ferrochelatase (heme synthetase) deficiency.** There is a defect in the ability of Fe^{++} to combine with protoporphyrin to produce heme. **Lead poisoning** acts at the same step, as an inhibitor of heme synthetase. Lead poisoning may thus result in an anemia. Lead also has other kinds of toxic effects (see #90). Lead can interfere with other kinds of proteins by interacting with sulfhydryl groups, causing more diffuse damage, particularly in the central nervous system. Lead resembles calcium in certain respects and can deposit in bone, where it may later be released, with delayed toxic effects. **Penicillamine** and EDTA bind to lead and may be used in treating lead intoxication.

#100. (D-13) **The hemoglobinopathies.** Mutations that change a single amino acid in a hemoglobin molecule may cause abnormally functioning red blood cells and anemia. Many hemoglobinopathies may be distinguished on hemoglobin electrophoresis. **Sickle cell anemia** is a classic example.

In sickle cell anemia, an abnormal hemoglobin is produced that results in sickle-shaped red cells, particularly under conditions of oxygen deprivation. These abnormally sticky cells may clog small blood vessels. The diagnosis may be made by examining the cells on a blood smear, and by hemoglobin electrophoresis. As in many hereditary conditions, homozygotes are far more severely affected. About 10% of American blacks are heterozygous (sickle cell trait) for sickle cell anemia, whereas about 0.4% are homozygous. Normal hemoglobin (HbA) has two alpha and two beta chains (alpha2,beta2). In sickle cell anemia, there is a change in the beta chain. Individuals with sickle cell trait have a mixture of HbA and HbS. Individuals homozygous for sickle cell have HbS but not HbA (fig. 10.2).

There are many kinds of amino acid substitutions that may produce abnormal hemoglobins besides that found in sickle cell anemia, but these conditions will not be elaborated upon here.

Genetic conditions that cause anemia may be due to other factors than a mutation causing a single amino acid change in the hemoglobin molecule. In certain **thalassemias,** for instance, the polypeptide chains are normal but abnormal quantities are produced. In **alpha thallassemia,** the alpha chain is underproduced. In **beta thallassemia,** the beta chain is underproduced. There may be a relative overabundance of other chains, such as beta or gamma chains. **Thallassemia minor** is the heterozygous type of beta thallassemia; **thallassemia major** is the homozygous type. One type of beta thallasemia appears to result from a gene mutation to form a stop codon. This stops beta-globin early in the formation of the protein chain. Thus, for all practical purposes, the chain doesn't even form. Certain types of alpha thallassemias may result from mutation and disappearance of a stop codon, resulting in abnormally long alpha-hemoglobin molecules.

In other anemias, the problem may not be the hemoglobin, but other factors that disrupt red cell energetics, as follows:

As red blood cells do not have mitochondria or a Krebs cycle, their energy is derived from glycolysis (ATP pro-

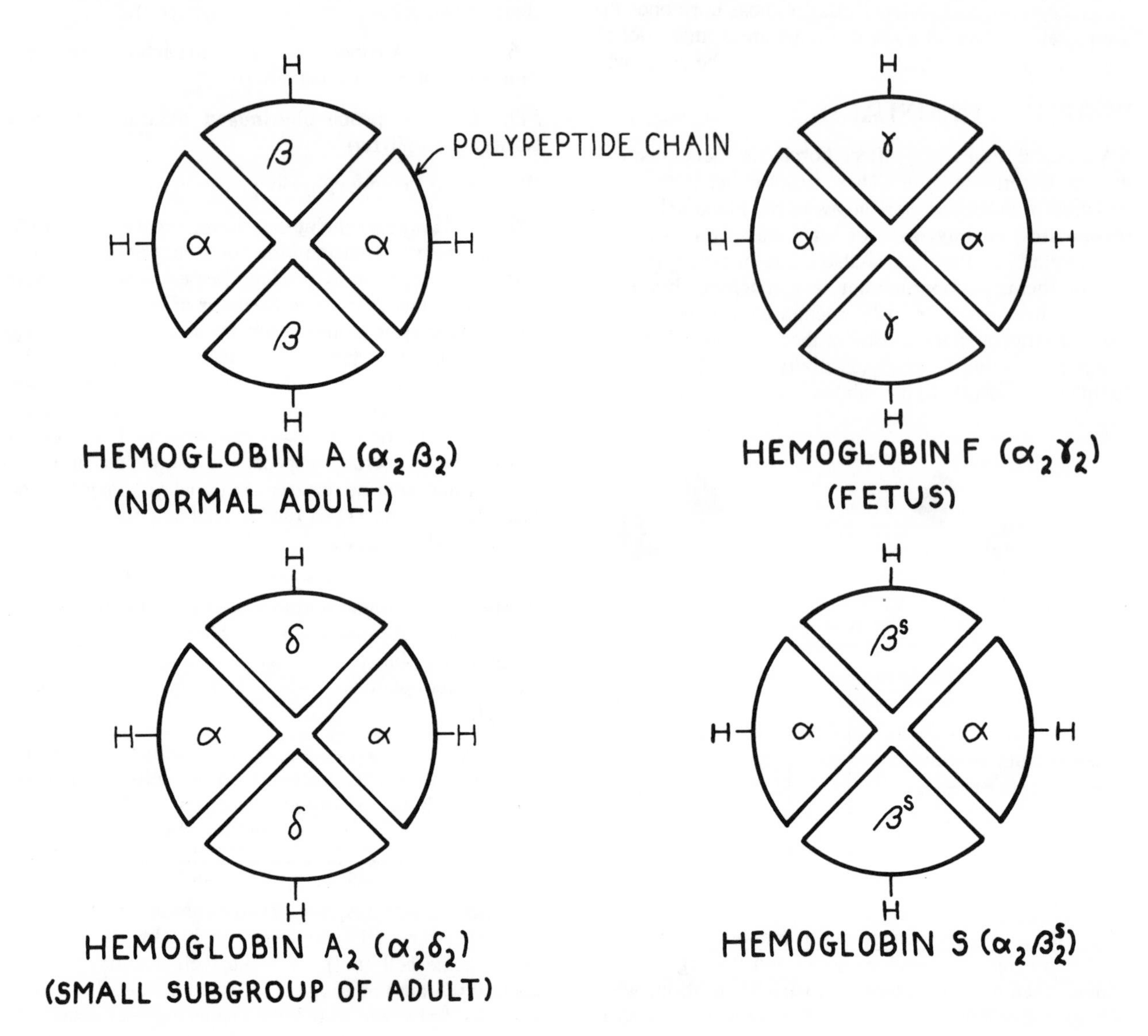

Fig. 10.2. Some hemoglobin types. H, hemoglobin.

duction) and from the HMP shunt (NADPH production). Defects either in glycolysis or in the HMP shunt may result in red cell abnormalities. Sometimes the defect centers in red blood cells. In other cases, it is more generalized, as glycolysis and the HMP shunt are generalized processes. Conditions that affect glycolysis or the HMP shunt include conditions #1–10).

Other disorders of red cells include problems of cell membrane functioning, e.g., **hereditary spherocytosis.**

Disorders of Bilirubin Metabolism

When red cells get old and are destroyed, mainly in the spleen, the heme ring is broken and the Fe^{++} is released, forming **biliverdin,** an open-chain molecule with no iron (C-14). Biliverdin changes to unconjugated ("indirect") bilirubin by reduction of its central carbon. The resultant unconjugated bilirubin molecule is carried to the liver by albumin (as unconjugated bilirubin is not water-soluble). Then:

a. Liver cells take up the **unconjugated bilirubin.**

b. They transform it into **conjugated bilirubin** by conjugating it with glucuronic acid (the resultant more polar molecule is now soluble in water).

c. The liver cells release the conjugated bilirubin into the **biliary duct system,** which carries it to the small intestine.

d. In the small intestine, the conjugated bilirubin is transformed by bacteria into **urobilinogen,** some of which is excreted and some of which is **reabsorbed,** carried back to the liver, and reexcreted into the bile. Urobilinogen may be detected in the urine when blood urobilinogen levels are high.

Defects may occur at any point in this sequence:

A. **Excessive breakdown of blood** (as in hemolytic anemias) can overwhelm the liver and result in elevation of the amount of circulating **unconjugated bilirubin.** This relatively non-polar molecule is not excreted in the urine but the secondary rise in the water-soluble urobilinogen may be detected in the urine as a laboratory test consistent with hemolytic anemia and certain liver disorders that involve poor urobilinogen reprocessing. Unconjugated (as well as conjugated) bilirubin, may be deposited in a number of tissues, causing jaundice. Unconjugated (lipid-soluble) bilirubin (but not conjugated, lipid-insoluble, bilirubin) can cross the blood brain barrier and cause toxic effects when deposited in the brain **(kernicterus),** particularly when deposited in the lipid-rich basal ganglia.

B. **Poor liver uptake** or poor conjugation of the unconjugated bilirubin. This likewise will result in increased serum unconjugated bilirubin. Decreased excretion of bilirubin in the bile will result in light-colored stools, as the by-product of urobilinogen metabolism, stercobilin, contributes to the normal brown color of stool.

C. **Inadequate release** of conjugated bilirubin into the bile, or overt bile duct obstruction. As the bilirubin is already conjugated, one may find a rise in the blood of conjugated (water soluble) bilirubin. As conjugated bilirubin can be excreted in the urine, the urine tests positive for bilirubin. Hepatitis and cirrhosis damage the liver and generally there is a rise in both unconjugated and conjugated bilirubin, more so of the conjugated form.

#101. (D-14) **Decrease in UDP-glucuronyl transferase activity** at this step. The liver normally conjugates "indirect" bilirubin to form conjugated ("direct") bilirubin which can be readily excreted through the bile. If such conjugation is impaired, as in deficiency of UDP-glucuronyl transferase, unconjugated bilirubin may accumulate in the blood, causing jaundice. This may occur transiently in newborns who have not as yet fully developed their UDP-glucuronyl transferase system. Usually, this is benign, but in severe cases, unconjugated (lipid soluble) bilirubin blood levels may be high enough to deposit in the brain and cause brain dysfunction **(kernicterus).**

Mild decreases in UDP-glucuronyl transferase are found in a variety of conditions. In **Gilbert's syndrome,** patients commonly are asymptomatic but may have mildly elevated unconjugated bilirubin. The enzyme defect is more severe in **Crigler-Najjar syndrome.** Phenobarbital stimulates the glucuronyl transferase activity and may be useful in reducing the jaundice. Phototherapy with white or blue light helps reduce neonatal jaundice by changing the unconjugated bilirubin to more water-soluble forms that can be excreted through the bile without necessitating conjugation.

The elevation of bilirubin that commonly occurs in newborns sometimes is not due to a deficiency in glucuronyl transferase but to blood group antigen-antibody incompatibilities which result in excess red cell destruction. Here, too, there is a rise in unconjugated bilirubin. Phototherapy and blood transfusion may both be helpful here.

#102. (D-14) **Drug detoxification in the liver.** UDP-glucuronate is important in the conjugation and detoxification of a number of drugs. The body has various ways of detoxifying drugs, and the liver is a particularly important site of such detoxification. A common strategy is to change the drug to a less toxic form, or to a more polar form that is soluble and excretable through the bile or urine. The kinds of reactions in which detoxification occurs include:

a. Conjugation

EXAMPLE: Conjugation with glucuronate (via the enzyme UDP-glucuronyl transferase) is the most important form of conjugation and is used not only for bilirubin but for deactivating many drugs (for instance, the antibiotic chloramphenical and the analgesic phenacetin). Conjugation may also occur through attachments other than glucuronide, such as methyl, acyl and sulfate groups.

b. Oxidation

EXAMPLE: ethanol + NAD^+ ⟶ acetaldehyde + NADH + H^+

c. Reduction

EXAMPLE: $R-CCl_3 \longrightarrow R-CHCl_2$ (as in the transformation of carbon tetrachloride or the anesthetic halothane)

d. Hydrolysis

EXAMPLE: $R-\overset{O}{\overset{\|}{C}}OCH_2-R' \xrightarrow{H_2O} R-\overset{O}{\overset{\|}{C}}OH + HO-CH_2-R'$
(acetylcholine and succinylcholine may be transformed in this way)

Newborns have relatively underdeveloped microsomal UDP-glucuronyl transferase. UDP-glucuronyl transferase detoxifies the antibiotic chloramphenicol. If newborns are given chloramphenicol, large amounts of the latter may accumulate in the body with toxic effect (the **grey baby syndrome**). Hence, chloramphenicol is avoided in newborns. On the other hand, phenobarbital stimulates the activity of the microsomal system and may be useful in treating the grey baby syndrome by increasing glucuronyl transferase activity. Phenobarbital will also stimulate the microsomal biotransformation of a variety of other drugs.

The combination of phenobarbital and alcohol is dangerous, apart from their additive sedative effects. Barbituates are detoxified by oxidation. Ethanol inhibits the reaction by being a competing substrate for oxidation, allowing the barbituate level to remain high.

#103. (E-14) Defective excretion of conjugated bilirubin into the bile. This occurs in the **Dubin-Johnson** and **Rotor syndromes.** The resultant hyperbilirubinemia is mainly of the conjugated type as is the hyperbilirubinemia of extrahepatic obstruction of the biliary ducts (as by tumor or gallstones).

THE INFIRMARY

The infirmary includes an inventory of vitamins, hormones, minerals, drugs, and laboratory tests. These are all scattered throughout Biochemistryland and are collected in the Infirmary for discussion as a whole.

Vitamins (indicated by green rectangles on the map) (see also pg. 41).

Water soluble vitamins (B_1, B_2, B_6, B_{12}, C, folacin, biotin, niacin, pantothenate) generally wash out of foods easily, and also wash out of the body relatively easily, (hence, are less easily stored in the body—an exception is vitamin B_{12} which is stored excellently, particularly in the liver). One may thus become depleted relatively quickly of most water soluble vitamins. Fortunately, so many foods are rich in them. Toxicity reactions, on the other hand, are more likely with fat-soluble vitamins, as they are so well-stored and are not eliminated easily from the body.

Deficiency of any of the vitamins may occur with inadequate intake. Fat-soluble vitamin deficiency may particularly occur in clinical conditions that affect fat absorption in the gut. Other intestinal wall diseases may affect vitamin absorption in general, causing deficiency of both water and fat soluble vitamins.

Vitamin D is partly produced by the skin on exposure to sunlight; vitamin K and biotin are produced in part by intestinal bacteria; and vitamin B_{12} is especially well-stored. Hence, it is uncommon to have a deficiency of these vitamins on the basis of dietary intake alone. Infants, who have less bacterial intestinal flora and relatively little stores of vitamin K have a greater need than adults for vitamin K in the diet.

Clinical points relevant to specific **water-soluble** vitamins are as follows:

#104. (F-9) **Vitamin B_1 (thiamine):**

COMMON FOOD SOURCES: wheat germ and whole grain cereals; fish, meat, eggs, milk, green vegetables

DEFICIENCY RESULTS IN: 1) **beri-beri.** This is associated with eating a diet of refined rice or with excess cooking of food, which decreases B_1 through washing out or through heating. The patient experiences cardiac and neurologic complications such as palpitations, edema, general weakness, pins-and-needles sensations and pain in the legs. 2) **Wernicke-Korsakoff syndrome.** This is correlated with alcoholism, but appears to be due to a nutritional deficiency rather than direct effects of alcohol. It may exist as a separate hereditary condition requiring large amounts of thiamine. There is a psychosis, with confabulation (fabrication of stories) and recent memory loss and other neurologic problems (nystagmus, eye muscle weakness, poor muscle coordination). Alcoholics undergoing delirium tremens (DT's) commonly receive intravenous thiamine as part of their in-hospital therapy.

#105. (F-9) **Vitamin B_2 (riboflavin):**

COMMON FOOD SOURCES: fish, meat, eggs, milk, green vegetables.

DEFICIENCY RESULTS IN: **angular stomatitis** (cracks in the corner of the mouth), inflammation of the tongue, seborrheic dermatitis, and anemia.

#106. (F-9) **Vitamin B_6 (pyridoxine; pyridoxal; pyridoxamine)**

COMMON FOOD SOURCES: whole grain cereals, fish, meat, eggs, green vegetables.

DEFICIENCY RESULTS IN: dermatitis, glossitis, anemia, and neurologic disturbances such as peripheral neuropathy and convulsions.

COMMENT: Deficiency is rare, as B_6 is so common in foods. Isoniazid, an antituberculosis drug, can interfere with B_6 and induce symptoms of B_6 deficiency. As pyridoxine is necessary in the steps that convert tryptophan to niacin (I-8), a deficiency of B_6 may result in niacin deficiency.

#107. (F-9) **Vitamin B_{12} (cyanocobolamin):**

COMMON FOOD SOURCES: Only microorganisms make B_{12} (not even plants make it). Large quantities are stored in the body, especially in the liver, enough to last 3 or more years, which is not the case for other water-soluble vitamins. We acquire vitamin B_{12} through ingesting meats (especially liver) and dairy products, including food fermented by bacteria, such as yogurt, soya sauce and sauerkraut.

DEFICIENCY: **Pernicious anemia.** Conceivably, one could get B_{12} deficiency on a purely vegetarian diet, but this is rare. Deficiency is more likely with diseases of the intestine that impede absorption (e.g., tropical sprue, regional enteritis). The tapeworm Diphyllobothrium latum may deplete B_{12} stores. A deficiency of gastric **intrinsic factor** (a glycoprotein) may result in B_{12} deficiency, as intrinsic factor is important in facilitating B_{12} absorption in the bowel. Intrinsic factor deficiency may occur following gastrectomy or as an entity in itself, in pernicious anemia. Intrinsic factor deficiency sometimes results from an autoimmune disease.

Patients with pernicious anemia have a macrocytic anemia, hypersegmented leukocytes, and neurosensory deficits (especially poor position sense) related to degeneration of the posterior columns of the spinal cord. Diagnostic clues include decreased B_{12} levels in the serum; a positive **Schilling test** in which radioactive labeled B_{12} is ingested, with subsequent assays of the degree of absorption; elevated urine levels of methylmalonate (D-11); decreased acidity of gastric secretion (as low stomach acidity is associated with lack of intrinsic factor and poor B_{12} absorption). Treatment of pernicious anemia consists of B_{12} injections to offset the poor intestinal absorption.

#108. (F-9) **Vitamin C (ascorbic acid):**

COMMON FOOD SOURCES: many fruits and vegetables.

DEFICIENCY: Man, other primates, and guinea pigs cannot synthesize vitamin C. Nor is it stored in the body. Hence, the need for continued replenishment, and the development of **scurvy** with lack of vitamin C. In scurvy, there are swollen gums and easy bruisability, coinciding with defective collagen formation. Serum levels of ascorbic acid may be tested for diagnosis.

EXCESS: There is concern about the possibility that certain susceptible individuals may develop renal stones or hemolytic anemia from megadoses of vitamin C.

#109. (F-9) **Folacin, (=folate=folic acid):**

COMMON FOOD SOURCES: cereals, liver, fruit, green leafy vegetables.

DEFICIENCY: Folate deficiency resembles B_{12} deficiency so far as the anemia goes, but without the neurologic abnormalities. Unlike B_{12}, for which there are tremendous body stores, folate needs continued replacement, and poor diet is the most common cause of folate deficiency. Serum levels of folate may help establish the diagnosis.

#110. (F-9) **Biotin:**

COMMON FOODS: widespread in foods; also produced by intestinal bacteria.

DEFICIENCY: Deficiency of biotin is rare except in people who only eat raw eggs, as egg whites contain **avidin,** a protein that inhibits biotin absorption. There may be a dermatitis.

#111. (F-9) (I-7) **Niacin** (=nicotinic acid). Nicotinamide is a related amide compound with similar activity as niacin):

COMMON FOODS: wheat germ, liver, fish, peanuts.

DEFICIENCY: **Pellagra.** Niacin may be produced from tryptophan (I-8). Niacin deficiency therefore is most likely in persons with low intake of both niacin and tryptophan. People who eat mainly **corn** may develop niacin deficiency as corn is low in tryptophan. In **pellagra,** the patient develops the 3 D's: **D**iarrhea, **D**ermatitis, and **D**ementia. Diagnostic testing is difficult and may best be done by seeing improvement with niacin ingestion.

#112. (F-9) **Pantothenate (=pantothenic acid):**

COMMON FOODS: present in many foods.

DEFICIENCY: Deficiency is very rare. The patient experiences fatigue, nausea, vomiting, abdominal pain, and neurosensory disturbances.

Clinical conditions associated with the **fat-soluble** vitamins (A,D,E,K) are as follows:

#113. (F-9) (F-11) **Vitamin A (retinol):**

COMMON FOODS: liver, fish, fortified milk, eggs, green leafy vegetables.

DEFICIENCY: Vitamin A deficiency results in night blindness and **xerophthalmia** (dry cornea and conjunctiva, sometimes with ulceration of the cornea). Non-ocular changes may also occur: dry skin and mucous membranes. Deficiency may result from poor dietary intake, or poor absorption, as from bowel disease, or a defect in bile flow that causes fat malabsorption. Poor protein intake may result in a reduced level of the transport protein that carries vitamin A in the blood stream.

TOXICITY: overdosage may result in vomiting, headache, depressed mental status, hair loss, skin peeling, and sometimes death.

#114. (F-9) (G-11) **Vitamin D:**

COMMON FOODS: fortified milk, fish liver oils, milk products, meat.

DEFICIENCY: Deficiency causes hypocalcemia and hypophosphatemia, with resultant **rickets** (bending with poor calcification of developing bone in children) or **osteomalacia** (decalcification and softening of bones in adults). In both of these there is impaired mineralization. These should be distinguished from another condition, **osteoporosis,** in which there is reduction in bone mass as a whole, rather than select reduction of the mineral content.

TOXICITY: Overdosage may result in hypercalcemia, and renal stones.

#115. (F-9) (G-10) **Vitamin E:**

COMMON FOODS: vegetable oils, green vegetables.

DEFICIENCY: Evidence is inconclusive as to what effects vitamin E deficiency may have in humans. Perhaps it may be associated with dystrophic changes in muscle and hemolysis of red blood cells.

TOXICITY: An excess may interfere with the action of various hormones, with vitamin K and blood clotting, and with white blood cell functioning.

#116. (F-9) (G-10) **Vitamin K:**

COMMON FOODS: widespread in animals and plants.

DEFICIENCY: Vitamin K is important in the blood clotting mechanism. Deficiency results in hemorrhages (especially in newborns—**hemorrhagic disease of the newborn**) and an elevated prothrombin time on laboratory testing.

TOXICITY: Hemolytic anemia; kernicterus (deposition of bilirubin in the brain) in newborns.

COMMENT: **Dicumarol** is a drug used for its anticoagulant effect. It looks like vitamin K and interferes with its functioning in blood clotting.

The Hormones (see also pg. 45)

#117. (C-5) **Thyroxine:**

DEFICIENCY: In children, deficiency results in a dwarf, commonly with an enlarged tongue and abdomen and mental retardation. In the adult, hypothyroidism results in **myxedema,** in which there are dry, edematous skin, mental and physical sluggishness, decreased tendon reflexes, decreased body metabolic rate and, sometimes, a goiter (enlarged thyroid gland). Hypothyroidism is most commonly due to underproduction of thyroxine by the thyroid gland.

EXCESS: results in **Graves' Disease,** in which there may be a goiter, exophthalmos (protrusion of one or both eyes), and hyperactivity, with increased body metabolic rate. Hyperthyroidism most commonly is due to overproduction of thyroxine by the thyroid gland.

#118. (G-13) **Glucocorticoids** (cortisol being most important):

EXCESS: results in **Cushing's syndrome.** Cortisol's effects, like those of other glucocorticoids, is generally opposite to that of insulin. It causes gluconeogenesis and breakdown of protein and fat. It also has antiinflammatory effects. In Cushing's syndrome there are high plasma cortisol levels, resulting in abnormal glucose metabolism (diabetes mellitus), abnormal fat distribution (obesity, buffalo hump, "moon face"), abnormal protein breakdown (osteoporosis, myopathy, purple stria on the abdomen) with a coinciding increase in androgen secretion (hirsutism, amenorrhea). Cortisol also has some degree of mineralocorticoid effect, and excess cortisol may result in hypertension and hypokalemia. Other effects include ulcers (there is stimulation of gastric acid and pepsin production), decreased inflammatory response, and psychosis. Excess of glucocorticoids may be due to an ACTH-secreting tumor.

#119. (G-12) **Mineralocorticoids** (aldosterone being most important):

DEFICIENCY: In **Addison's disease,** there is adrenal cortical insufficiency of both glucocorticoids and mineralocorticoids. The patient may have hypotension, low serum sodium, weakness, and excess skin pigmentation. The condition may be secondary to autoimmune disease affecting the adrenal gland, or infection.

EXCESS: results in **Conn's syndrome,** in which the patient has hypertension and low serum potassium. This may be secondary to a hormone-producing tumor.

#120. (C-5) **Epinephrine:**

EXCESS: results classically from a hormone-producing tumor **(pheochromocytoma).** The patient experiences episodes of hypertension, cardiac palpitations, and anxiety. Urinary VMA may be elevated.

#121. (H-13) **Androgens** (testosterone being most important):

DEFICIENCY: in utero, results in failure of male genital development. Deficiency in childhood results in failure to develop secondary male characteristics. This may be due to castration or a hypofunctioning hypothalamus. In **male pseudohermaphroditism,** there appears to be a defect in the cellular recognition proteins for androgens, so that androgens, while present, do not stimulate the formation of male characteristics.

EXCESS: in childhood results in precocious puberty; may result from a hormone-secreting tumor. In adults, testosterone administration has been used by athletes to increase body mass, but may be associated with side effects, including cardiac and liver disease.

#122. (H-13) **Estrogens** (the most important of which is estradiol):

DEFICIENCY: in childhood (e.g., in Turner's syndrome) results in failure to develop secondary female sex characteristics. In the adult, loss of menstruation results, as may occur following menopause or removal of the ovaries.

EXCESS: in childhood may occur with a hormone-producing tumor, resulting in precocious puberty.

Other hormones, not listed on the map but associated with clinical conditions include:

Insulin:

DEFICIENCY: results in diabetes mellitus, the hallmarks of which are hyperglycemia, glycosuria, and in some cases ketoacidosis. Excessively high blood sugars may result in osmotic fluid loss and coma. Chronic diabetes is often associated with vascular disease. Diabetes may be secondary to underproduction of insulin or to the production of an abnormal insulin. In such cases, insulin may be effectively administered as treatment. In other cases, insulin production is normal but there is a defect in the cell receptor for insulin or in cellular events beyond the receptor step. Such cases may be resistant to treatment with insulin.

EXCESS: results in hypoglycemia and, in some cases, coma. This most commonly occurs with excess insulin administration, but may occur with an insulin-secreting tumor.

Glucagon:

EXCESS: may occur in a glucagon secreting tumor and may, among other things, result in life-threatening hyperglycemia.

Growth Hormone:

DEFICIENCY: in childhood results in a normally proportioned midget.

EXCESS: results in a normally proportioned giant when overproduction occurs in childhood. Overproduction in an adult results in **acromegaly.** Such individuals have abnormally proportioned growth features, especially enlarged hands, feet and jaw. There may be hyperglycemia, as growth hormone is also involved in glucose metabolism.

Antidiuretic hormone (ADH; vasopressin):

DEFICIENCY: results in **diabetes insipidus,** in which the patient may excrete large volumes of urine and ingest large volumes of water.

EXCESS: results in water retention and resultant hyponatremia. ADH may be inappropriately secreted by certain tumors and, in response to certain inflammations, by normal lung or pituitary tissue.

Calcitonin:

EXCESS: is found in **medullary thyroid carcinoma** and may result in a mild hypocalcemia.

Parathyroid hormone:

DEFICIENCY: results in hypocalcemia and tetany (spasm on flexion of the wrist and joints).

EXCESS: results in hypercalcemia; calcium may deposit in various tissues or precipitate as renal stones; there is bone demineralization.

ACTH, TSH, LH, FSH, and RENIN: Deficiencies and excesses are reflected in changes in activity of their respective target organs.

Prolactin:

EXCESS: galactorrhea (excess lactation); may result rarely from a hormone-secreting tumor.

Gastrin:

EXCESS: results in gastric hyperacidity and ulcers; occurs with gastrin-producing tumors **(Zollinger-Ellison syndrome).**

Minerals

#123. (F-9) **Minerals** are very important in Biochemistryland, but have not been emphasized, largely because their functions are so closely intertwined with organ physiology, a topic beyond the province of this book and customarily dealt with in physiology courses. Nonetheless, a brief overview of the major and trace minerals is presented below:

SODIUM: The most abundant extracellular cation (positive ion). It influences the degree of water retention in the body and is an important participant in the control of acid-base balance.
Deficiency—may result in neuromuscular dysfunction.
Excess—may result in hypertension and fluid retention.

POTASSIUM: The main intracellular cation. It is important in all cell functioning including myocardial depolarization and contraction.
Deficiency—may result in neuromuscular dysfunction.
Excess—may cause myocardial dysfunction.

CHLORIDE: An important anion (negative ion) in the maintenance of fluid and electrolyte balance; an important component of gastric juice.

CALCIUM: The most abundant of the body's minerals; an important component of bones and teeth; important participant in regulation of many metabolic processes. When bound to the receptor protein **calmodulin,** calcium helps modulate the activities of many enzymes. Calcium is important in regulation of blood clotting, neural and muscular activity, cell motility, hormone actions, and other activities.

Deficiency—associated with rickets and osteomalacia, tetany (muscle spasm, especially in the wrists and ankles), and other neuromuscular problems.

Excess—causes hypercalcemia and renal stones.

PHOSPHORUS: Apart from being an important component of bone, phosphorus, especially in the form of the phosphate molecule, is universally important in the structure and functioning of all cells.

Deficiency—associated with rickets in children, and osteomalacia in adults. Other defects occur in the functioning of red and white blood cells, platelets, and the liver.

MAGNESIUM: Important participant in reactions that involve ATP.

Deficiency—associated with metabolic and neurologic dysfunction.

Excess—associated with central nervous system toxicity.

Trace Elements

CHROMIUM: Enhances the effect of insulin.

Deficiency—results in defective glucose metabolism.

Excess—occurs in chronic inhalation of chromium dust and may lead to carcinoma of the lung.

COBALT: Part of the B_{12} molecule.

Excess—may result in gastrointestinal distress and neurologic and cardiac dysfunction.

COPPER: A part of a number of enzymes, including cytochrome oxidase.

Deficiency—may result in anemia and mental retardation.

Excess—results in liver disease, various neurologic disturbances, dementia, copper cataracts. These occur in **Wilson's Disease,** where there is excess copper deposition in the brain, liver, cornea, lens, and kidney.

FLUORIDE: Contributes to the hardness of bones and teeth.

Deficiency—associated with dental caries.

Excess—associated with stained teeth, nausea and other gastrointestinal disturbances, and tetany.

IODIDE: Part of the hormone thyroxine.

Deficiency—results in hypothyroidism.

Excess—results in hyperthyroidism.

IRON: Its most important functions are its inclusion in the hemoglobin molecule, certain enzymes, and the cytochromes.

Deficiency—anemia.

Excess—**hemochromatosis** (abnormal iron deposits and damage to liver, pancreas and other tissues).

MANGANESE: Needed to activate a variety of enzymes, including enzymes involved in the synthesis of glycoproteins, proteoglycans and oligosaccharides.

Deficiency—may result in underproduction of the latter molecules.

Excess—may result in Parkinson-like symptoms (shaking, slowness, stiffness) and psychosis.

MOLYBDENUM: An important component of certain enzymes (e.g., xanthine oxidase).

NICKEL: May stabilize the structure of nucleic acids and cell membranes.

Excess—may be associated with carcinoma of the lung.

SELENIUM: Part of the enzyme glutathione peroxidase, which, like vitamin E, acts as an antioxidant.

Deficiency—may result in congestive heart failure.

Excess—causes "garlic" breath body odor and skeletal muscle degeneration.

SILICON: Associated with many mucopolysaccharides and may be important in the structuring of connective tissue. Excess, as in inhaling silica particles, may result in pulmonary inflammation **(silicosis).**

ZINC: A component of many enzymes, including lactate dehydrogenase and alkaline phosphatase.

Deficiency—associated with many problems, including poor wound healing, hypogonadism, decreased taste and smell, and poor growth.

Excess—associated with vomiting from gastrointestinal irritation.

Antibiotics

Antibiotic effectiveness depends on the drug's differential toxicity to microorganisms, as opposed to humans. One way in which an antibiotic may act against bacteria is by inhibiting cell wall synthesis, as bacteria have cell walls, whereas human cells do not. The cell wall contains glycoproteins that are unique to bacteria, and antibiotics may affect the synthesis of these cell wall components. As gram positive and gram negative bacteria differ in their cell wall structure (fig. 10.3), certain antibiotics are more specific for one or the other of these types of bacteria. In general, antibiotics act by one of the following mechanisms:

a. Inhibiting cell wall synthesis (e.g., the penicillins). Many gram negative bacteria are resistant to penicillin, as the outer membrane of gram negative bacteria blocks the entrance of penicillin. Bacteria may also become resistant to penicillin by producing penicillinase, an enzyme that breaks the beta-lactam ring of the penicillin molecule. Variants of penicillin have been developed that are effective even in the presence of penicillinase.

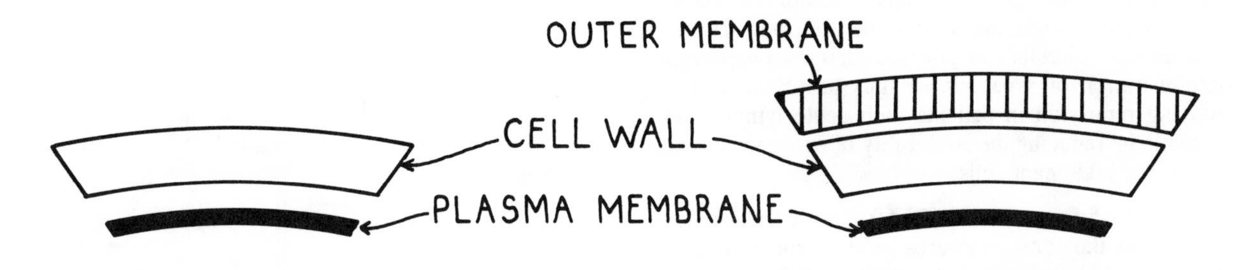

Fig. 10.3. The different surface structures of gram negative and gram positive bacteria.

b. Acting on the bacterial cytoplasmic membrane (e.g., the polymixin antibiotics).

c. Interfering with protein synthesis in bacteria (e.g., erythromycin and the tetracyclines).

d. Interfering with other aspects of cytoplasmic metabolism. For instance, sulfonamides interfere with the bacterial incorporation of PABA (para-aminobenzoic acid) in the formation of folic acid:

#124. (A-6) **Erythromycin and the tetracyclines.** These antibiotics act in part by interfering with bacterial protein synthesis.

#125. (A-2) **Sulfonamides.** Bacteria need PABA in order to form folic acid. Humans do not need PABA, as they get folic acid from the diet. Sulfonamides are competitive antagonists of PABA and thus can affect bacteria while not harming human cells (fig. 10.4).

#126. (A-2) It is common to combine **trimethoprim with sulfonamide** in the treatment of certain infections. The sulfonamide acts on bacteria to inhibit folate formation in bacteria by competitively inhibiting PABA incorporation (fig. 10.4). The trimethoprim acts on the dihydrofolate reductase step (between 7,8 dihydrofolate and 5,6,7,8 THF). Apparently, trimethoprim has a more damaging effect on bacteria and protozoans than it has on humans.

#127. (A-6) **Adenine arabinoside** is an antiviral agent that is an analogue of the purine adenine. It acts on DNA viruses by inhibiting viral DNA polymerase. **Idoxyuridine** also acts on DNA viruses. As a derivative of the pyrimidine deoxyuridine, it becomes incorporated into DNA and blocks viral multiplication. **Acyclovir,** a purine nucleoside analogue of guanine, converts to a triphosphate form which interferes with herpes simplex viral DNA polymerase. **Azidothymidine (AZT),** an antiviral agent used in AIDS therapy, is a thymidine analogue.

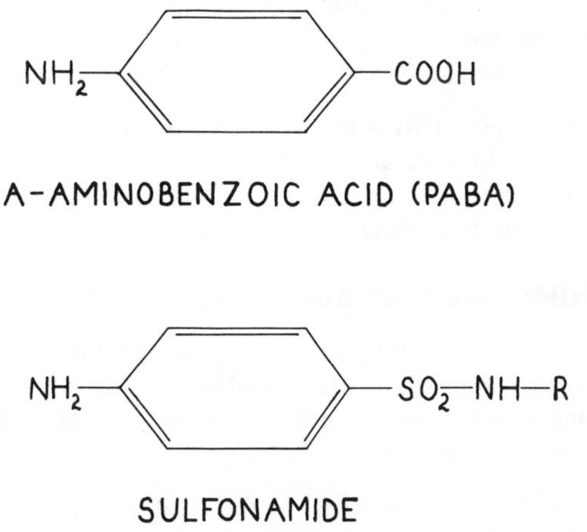

NH_2—(benzene ring)—COOH

PARA-AMINOBENZOIC ACID (PABA)

NH_2—(benzene ring)—SO_2—NH—R

SULFONAMIDE

Fig. 10.4. Similarity in structure between PABA and sulfonamide. Sulfonamides are competitive antagonists of PABA.

Antitumor Drugs

The strategy of antitumor agents is to kill tumor cells in preference to normal cells. A number of them act by inhibiting the formation of nucleotides or altering the structure of DNA. Others act in diverse ways:

#128. (A-2) (B-6) **Methotrexate** is an antitumor agent that is a folic acid antagonist. It looks like folic acid and inhibits the transformation of 7,8-dihydrofolate to tetrahydrofolate (THF). The rationale for using methotrexate as an antitumor agent is that THF is necessary for the transformation of dUMP to TMP (B-7) in pyrimidine biosynthesis and is also necessary for purine ring biosynthesis. Inhibiting these steps interferes with the formation of DNA and, hence, inhibits tumor cell division.

#129. (C-8) **Asparaginase** therapy of tumors. Certain tumors require exogenous asparagine for their survival, whereas normal cells can produce their own asparagine. Administration of asparaginase thus has been useful in inducing remission in patients with acute lymphoblastic leukemia by reducing the availability of exogenous asparagine to the malignant cells.

#130. (B-7) **5-fluorouracil (5-FU)** is an antitumor drug that acts at the same map area as does methotrexate in pyrimidine biosynthesis, but acts by a different mechanism. Through its resemblance to uracil, it inhibits thymidilate synthetase at this step and prevents the formation of thymidine, a pyrimidine. In a similar vein, **cytarabine** is an antitumor agent that acts through its resemblance to cytidine, another pyrimidine.

#131. (B-4) **Mercaptopurine** is an antitumor drug that looks like the purine adenine and acts partly by interfering with its incorporation into nucleic acids.

#132. (B-5) **Thioguanine** is an antitumor agent that resembles the purine guanine and acts by abnormally substituting for guanine in RNA and DNA.

#133. (A-6) **Alkylating agents** (e.g., nitrogen mustards, nitrosoureas) are antitumor agents that act by replacing a hydrogen in a DNA molecule with an alkyl group, resulting in the malfunctioning of DNA.

Other Drugs in Biochemistryland

While not wishing to tread too deeply into the field of pharmacology, there are a number of drugs whose actions relate to very specific areas of Biochemistryland. Some of them are as follows:

#134. (B-5) **Allopurinal** in the treatment of gout. Xanthine oxidase facilitates steps that lead to the production of urate. Allopurinal, which looks like hypoxanthine (fig. 10.5), is a competitive inhibitor of xanthine oxidase and thereby reduces urate production. (Xanthine oxidase normally functions at steps between hypoxanthine and xanthine, and between xanthine and uric acid).

#135. (F-1) **Hyaluronidase** cleaves hyaluronic acid. It is produced by certain bacteria and may facilitate their spread through tissue planes. Sometimes hyaluronidase is administered with drug injections to facilitate the spread of the drug through the injected tissues.

#136. (F-1) **Heparin** interferes with the activity of a number of clotting factors in the clotting cascade. It is a useful anticoagulant drug (see also #151).

#137. (D-7) **Phenformin** is a drug that prevents lactic acid from undergoing gluconeogenesis in the liver. It has been used as an antidiabetic agent, but has largely been discontinued as it tends to cause lactic acidosis. The **sulfonoureas** are other antidiabetic drugs that act partly by stimulating insulin release and partly by increasing the number of insulin receptors in peripheral tissues.

HYPOXANTHINE

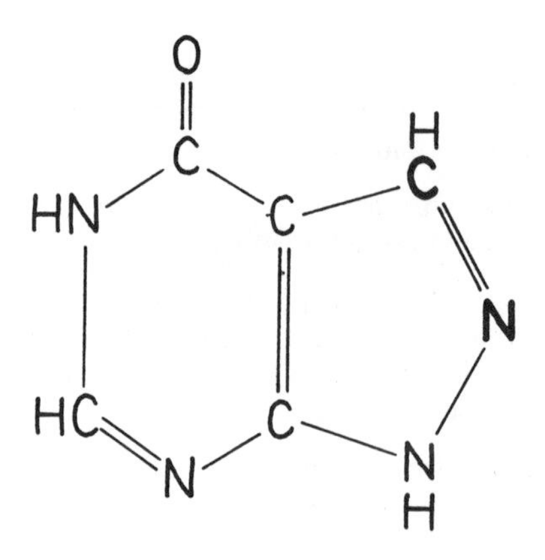

ALLOPURINOL

Fig. 10.5. Resemblance of hypoxanthine and allopurinol.

#138. (E-8) **Antabuse** is a drug used sometimes as a treatment for alcoholism. Antabuse prevents acetaldehyde from changing to acetate. This results in a marked buildup of acetaldehyde in the blood, making the patient feel quite sick when alcohol is ingested. **Metronidazole** is an antibiotic having a similar action, and patients must be cautioned not to drink alcohol while taking this drug.

#139. (I-10) **Monoamine oxidase (MAO) inhibitors.** Drugs that inhibit monoamine oxidase are used as antidepressants in certain patients. By inhibiting MAO at this step, there is an accumulation of serotonin and other amines (e.g. epinephrine), with resultant mood elevation. One must be cautious in administering certain other drugs with MAO inhibitors, as MAO inhibitors also inhibit liver oxidative enzymes that normally detoxify many drugs. The patient must be careful not to ingest foods high in tyramine (like cheese). Tyramine, a sympathetic-acting molecule, normally does not get into the general circula-

metabolized by MAO in the intestine and liver. MAO inhibitors allow it to circulate and can cause severe hypertension and possibly cardiac arrhythmias and stroke.

#140. (H-4) **Aspirin.** Aspirin decreases inflammation, pain, and fever. It also acts as an anticoagulant, by inhibiting platelet aggregation. Aspirin inhibits cyclooxygenase, an enzyme important in prostaglandin and thromboxane synthesis (fig. 4.4). **Indomethacin** and **phenylbutazone,** like aspirin, are non-steroidal antiinflammatory agents that inhibit cyclooxygenase. **Steroidal** antiinflammatory agents appear to act at a preceding step, inhibiting phospholipase A_2 activity and thus interfering with the release of arachidonic acid from its phospholipid (fig. 4.4).

Fish oils have been found to contain special polyunsaturated fatty acids, which in certain respects look like arachidonic acid. These **"omega-3"** fatty acids are so called because they have a C-C bond between the third and fourth carbons from the omega end (non-carboxylic end) of the fatty acid. They are converted to prostaglandin-like compounds which appear to have an effect in lowering blood pressure, decreasing platelet aggregation, and, to some degree, increasing the HDL/LDL ratio. There is considerable interest in their potential usefulness in the treatment and prophylaxis of hypertension and cardiovascular disease.

#141. (C-6) **Thiouracil** is used in the treatment of hyperthyroidism. Thiouracil drugs inhibit the coupling of the iodotyrosines to form T_3 and T_4.

Selected Laboratory Tests in Biochemistryland

#142. (B-9) **BUN (blood urea nitrogen).** The BUN may be elevated in renal disease (faulty renal excretion), as well as in states of dehydration. The BUN can be a useful index of renal function.

#143. (B-10) **Creatine phosphokinase (CPK).** Creatine phosphate (=phosphocreatine) acts as a reserve energy store, supplying phosphate to change ADP to ATP when needed. Physicians don't care too much about this, but do care about leakage of creatine phosphokinase (creatine kinase; CPK) as a detectable marker of cell injury. CPK comes as 3 isozymes (enzyme variants): MM, MB, and BB. MM is found mainly in skeletal muscle; MM and MB are found mainly in heart muscle; and BB is found mainly in the brain. Elevation of MB is suggestive of a myocardial infarction. Elevation of only serum MM occurs in muscle damage, e.g., following trauma.

#144. (B-10) **Creatinine** as a test of renal function. Creatinine, which is derived from creatine, normally is excreted almost totally by the kidney. Blood levels of serum creatinine, as well as urea, are useful indices of renal function, their elevation often being a sign of renal insufficiency.

#145. (E-11) **SGOT** (serum glutamate oxaloacetate transaminase), more recently called **AST** (aspartate aminotransferase), acts at this step. Both names make sense, depending on which way you read the chemical reaction. The enzyme is found in many areas of the body, but is most useful as a marker of liver or cardiac injury. It leaks out of the damaged cell and increases in the serum after myocardial infarction and liver injury (for instance, hepatitis) and may provide a clue as to the existence of these conditions.

#146. (E-7) **SGPT** (serum glutamate pyruvate transaminase), more recently called **ALT** (alanine aminotransferase), acts at this step. Both names make sense, depending on which way you read the chemical reaction). This enzyme is especially concentrated in the liver; it leaks out of the liver cell and rises in the serum with liver damage, as in hepatitis and mononucleosis. It does not significantly increase in myocardial infarction and, hence, the test is more specific than SGOT for liver disease.

GPT (ALT) is localized in the cytosol. GOT (AST) localizes to **both** cytosol and mitochondria and is more abundant in liver cells than SGPT. In conditions that damage the **entire** liver (e.g., cirrhosis, hypoxia, tumor) SGOT levels are relatively higher than SGPT levels. However, in liver conditions that mainly affect the cell membrane (e.g. viral hepatitis) SGPT levels are relatively higher, as the cytosol enzymes, rather than mitochondrial enzymes leak out.

Other enzymes are also useful indices of liver pathology. **Serum alkaline phosphatase** is often a useful indicator of liver and bone disease. The alkaline phosphatases are a diverse group of enzymes that catalyze reactions in which a phosphate is removed from a phosphate ester, especially at an alkaline pH. Physicians don't care about this. They do care that serum alkaline phosphatase levels often rise with bone breakdown (as in tumor infiltration) and in liver disease, especially where there is obstruction of the bile duct. **Acid phosphatase** is particularly rich in the prostate. A rise in its serum levels often provides a clue as to the presence of prostatic carcinoma.

#147. (D-7) **Serum LDH (lactate dehydrogenase) elevation.** LDH is a widespread intracellular enzyme that acts at this step. Detection of high serum levels is rather nonspecific for localizing the site of damage, but its measurement is helpful in confirming myocardial infarction or injury to the liver, skeletal muscle, or certain other tissues. The presence of 5 different isozymes of LDH helps to further localize the injury, as they are fairly tissue specific. For instance, LD_1 and LD_2 are elevated in myocardial infarction; LD_2 and LD_3 elevation occur in acute leukemia; LD_5 elevation follows liver or skeletal muscle injury.

#148. (F-9) **Sorbitol dehydrogenase.** Elevation of serum levels is a useful sign of liver cell damage.

The **A/G (albumin/globulin) ratio** is another useful index of liver dysfunction. Albumin is produced by the liver, whereas gamma globulin is produced by cells of the immune system. Thus, the A/G ratio may be decreased in liver disease.

#149. (D-1) **Amylase** testing. Amylase is a digestive enzyme that is produced mainly by the salivary glands and pancreas. It may be elevated in the serum in mumps or in pancreatitis, but sometimes in other conditions (e.g. peptic ulcer, intestinal obstruction, gallstones).

#150. (E-2) **Clinitest tablets** and **Clinistix testing** of urinary sugar. Monosaccharides are examples of reducing sugars. Reducing sugars reduce oxidizing agents such as the cupric ion ($Cu^{++} \rightarrow Cu^{+}$). Among the disaccharides, lactose and maltose are also reducing sugars, but sucrose (table sugar) is not. **Clinitest tablets** are used to test for reducing sugars in the urine and will detect a **broad range** of reducing compounds that can appear in the urine, including glucose (as in diabetes), glucuronate (which conjugates with many drugs and is excreted in the urine), fructose (as in genetic fructosuria), galactose (genetic galactosemia), lactose (in late pregnancy and lactation, and in congenital lactase deficiency), the pentoses (genetic pentosuria), and homogentisic acid (alkaptonuria). Clinitest tablets will not, however, detect sucrose. This fact has led to the failure of individuals to deceive the physician into thinking they have diabetes (e.g., to avoid military induction) by pouring table sugar into their urine.

Clinistix dipsticks are more **specific** than clinitest tablets for detecting glucose, as their reaction depends on the fact that they have **glucose oxidase. Ascorbic acid** can cause false negative results with Clinistix. Clinitest tablets, which react with many reducing substances, are useful as a general screen for reducing substances, especially in neonates, where the test may detect a variety of inborn errors of metabolism (e.g., galactosemia, fructosuria, congenital lactase deficiency, pentosuria, alkaptonuria).

#151. (G-11) **Coumarin, vitamin K, and heparin. Vitamin K** is necessary for the **synthesis** of a number of clotting factors produced by the liver, including prothrombin. The **coumarin** anticoagulants resemble vitamin K in structure and compete, preventing vitamin K from functioning in the liver, thereby decreasing the tendency to form prothrombin. The onset of coumarin anticoagulant action is relatively slow, requiring several days for the previously formed prothrombin to clear. The plasma **prothrombin time** is a useful laboratory test to assess the level of prothrombin and the effectiveness of coumarin anticoagulation. An elevated prothrombin time (signifying low prothrombin levels) may also occur in liver disease, as the liver produces prothrombin. **Heparin** is a different anticoagulant that has a more rapid onset of action as its action is more direct-interference with a number of steps in the clotting cascade, as well as reducing platelet adhesiveness. Heparin acts as an anticoagulant both in vivo and in vitro (e.g., in preventing a blood sample from clotting), in contrast to coumarin, which only acts in vivo.

CHAPTER 11. BEYOND BIOCHEMISTRYLAND

Imagine the Biochemistryland map lying flat on the table. There are other maps in medicine that may be stacked on top of Biochemistryland at increasingly higher hierarchical levels (fig. 11.1). These include Cell Biologyland, Histologyland, Anatomyland, and the realm of the Individual and Society. At the bottom, Biochemistryland, there are chemical reactions that interconnect extensively with one another. The chemical reaction arrows at this level are not only the arrows on our own Biochemistryland map but all the additional feedback arrows from reaction products, hormones, and other molecules not shown on the map. Biochemistryland really resembles a giant spider web, or system of rivers, in which motion at any point can cause changes throughout much of the system.

There are also connections within each of the higher level maps. Changes within organelles can affect other organelles in Cell Biologyland: defective mitochondria, for instance, may affect other organelle functions. In Histologyland, changes of certain cell types can affect the function of other cell types: for example, a loss of substantia nigra nerve cells in the brain stem may lead to the degeneration of basal ganglia cells on which they normally connect. In Anatomyland, malfunction of one organ system can affect others: inadequate pulmonary function will affect the functioning of the heart and other organs. Interactions between individuals can affect societal functioning.

It is also true that each of the hierarachical levels interacts with one other. For instance, abnormalities of biochemical reactions can affect the functioning of select organs; hypoglycemia, for instance, may result in coma. Emotional stress can precipitate heart attacks, gastric hyperacidity, epinephrine production, etc., inducing changes all the way down to the biochemical level. Conversely, mood-altering drugs that act at the biochemical level can affect the functioning of the individual and the individual's interactions with society.

In medicine, we unknowingly continually shift between levels in working with different clinical conditions. We have seen a number of diseases scattered through Biochemistryland, but these are for the most part rare. Even the next higher level, Cell Biologyland, has relatively little clinically when compared with higher levels; for instance, there are myopathies that affect mitochondria, and there are lysosomal storage diseases, but these are uncommon. Clinical thinking in large part centers on higher levels. Histologyland has a great deal of tumor pathology. Anatomyland has an enormous bulk of clinical disorders related to organ pathology—the heart attacks, fractures, muscle spasms, strokes, liver and kidney diseases, etc. Levels of the individual and society contain many psychiatric and sociopathic disorders.

Thinking at different levels brings out additional information; new concepts of function arise on ascending through the hierarchy, just as a particular arrangement of bricks becomes something new—a "house". The concept of an organ is not ingrained in any particular reaction in Biochemistryland but in the integration of many complex reactions that forms something new, an organ. If my car

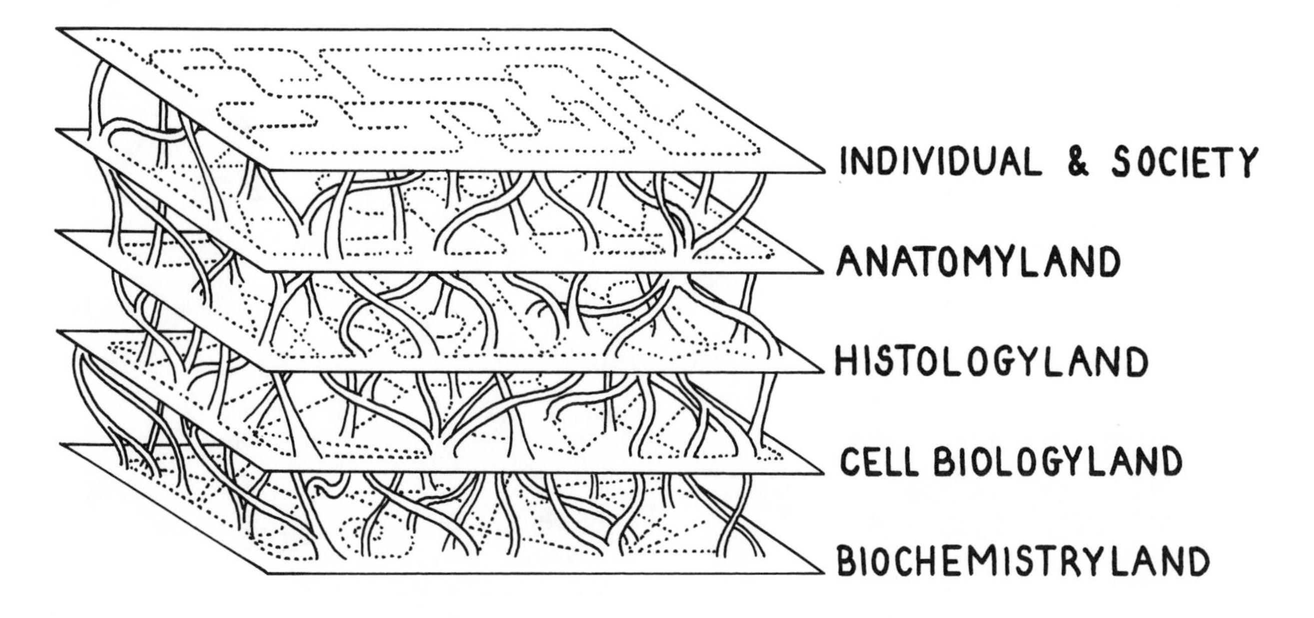

Fig. 11.1. The interconnection of Biochemistryland with other lands.

door gets smashed in, I could go to the body shop and ask for a particular combination of metal alloys, arranged as so many kinds of molecules packed together into a space of such and such dimensions. More appropriately, however, I will simply ask for a new car door, the "door" being the unit to focus on rather than the molecules that compose the door. Similarly, particular clinical disorders are best thought of as occurring at certain levels; when one deals with a lacerated liver, one is not going to think of the replacement and apposition of certain kinds and number of molecules that constitute liver cells; one is going to think in terms of surgical techniques that repair a lacerated liver. Even though everything can ultimately reduce itself to biochemistry, working at a particular hierarchical level can provide a better way of thinking about the same material.

While it is often more practical to think at a particular heirarchical level it is important not to get so fixated at any one particular level as to ignore the effects one level may have on another, as if they were separate and unrelated realms. One must remember that the individual levels interconnect. Higher level personal-social situations can affect one's biochemistry, and biochemical changes can affect one's personal-social status. Things like bedside manner, hope, and the placebo effect can have real effects on underlying biochemistry, just as biochemical changes can affect mood and behavior.

It is better, where possible, to treat the root cause of an illness rather than a peripheral manifestation of the illness. When one treats just the manifestation, one may eliminate it, but this does not eliminate the many other widespread changes that the original illness may have caused throughout the heirarchy of lands. If a depression is situationally induced, it is better, where possible, to correct it at the higher situational level than to deal with it through medication at the lower biochemical level; the originating problem otherwise persists and there may be medication side effects. Thoughts such as these may appear obvious. No one would think of trying to treat a heart attack by lowering elevated blood levels of LDH enzyme that leaked out of the damaged cardiac cells. In other situations the logic is less clear. Thus, there have been and still are controversies over whether lowering blood levels of cholesterol in hypercholesterolemia, or glucose in diabetes has a significant effect in preventing atherosclerosis. Do such clinical measures get at root causes, and if they do not, do such measures significantly affect the course of disease?

There are many medical conditions for which a root cause is unknown and one has to focus on treating peripheral manifestations (e.g., treating inflammation in arthritis with antiinflammatory drugs). Even when a root cause is known it may be difficult to treat if the problem is localized to a particular area of the body; the medication may not only affect the problem area in the body but have side effects on other regions—for example, the side effects of antitumor agents on normal cells. When the root cause is outside the body, e.g., smoking as a cause of lung cancer, elimination of the root cause can be both feasible and clinically effective. Hence, the importance of Preventive Medicine as an important influence on the hierarchical schema.

APPENDIX I. ISOMERS

Isomers are different chemicals that have the same molecular formulae.

1. **Chain isomers**—isomers that contain different molecular sequences within the chain:

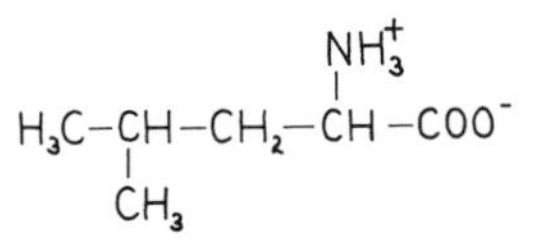

$H_3C-CH(CH_3)-CH_2-CH(NH_3^+)-COO^-$

LEUCINE

$H_3C-CH_2-CH(CH_3)-CH(NH_3^+)-COO^-$

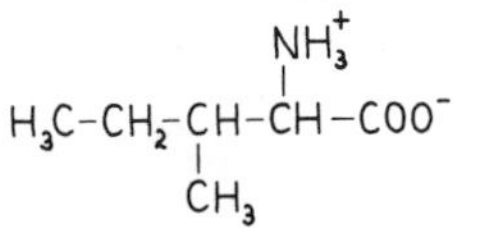

ISOLEUCINE

2. **Tautomers**—isomers that contain different bond sequences that fluctuate in equilibrium from one form to another:

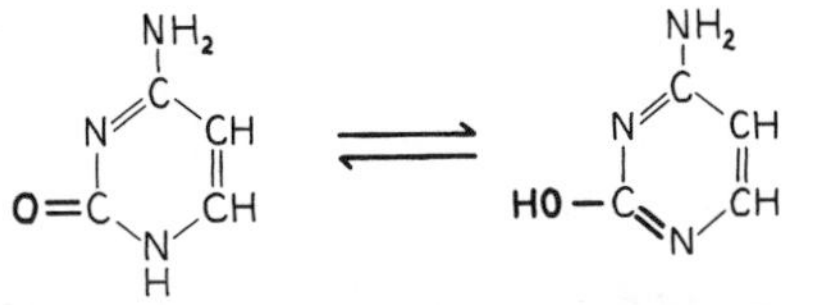

CYTOSINE (Lactam tautomer) ⇌ CYTOSINE (Lactim tautomer)

3. **Stereoisomers**—isomers that differ in their orientation in space. The terms (+) and (−) refer to the direction of optical rotation when exposed to polarized light.

A. Enantiomers—stereoisomers that are mirror images of one another:

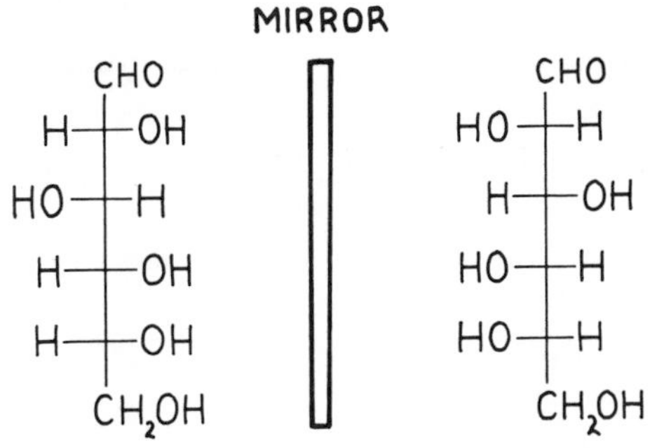

MIRROR

D-GLUCOSE		L-GLUCOSE
CHO		CHO
H–C–OH		HO–C–H
HO–C–H		H–C–OH
H–C–OH		HO–C–H
H–C–OH		HO–C–H
CH_2OH		CH_2OH

The terms "D" and "L" (dexter-right and levo-left) refer to the two mirror image forms that an enantiomer may have and in Biochemistryland generally refer to carbohydrates and amino acids. In most cases it is the "D" form of carbohydrates and the "L" form of amino acids that are utilized by the body.

B. Diastereomers—stereoisomers that are not mirror images of one another. **Epimers** are diasteriomers that differ in conformation around only one carbon atom.

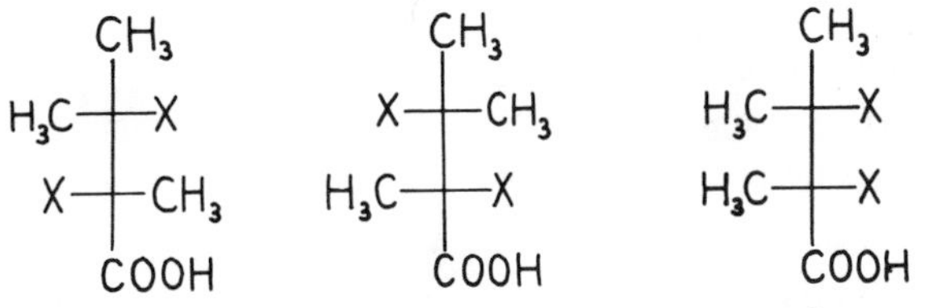

A	B	C
CH_3	CH_3	CH_3
H_3C–C–X	X–C–CH_3	H_3C–C–X
X–C–CH_3	H_3C–C–X	H_3C–C–X
COOH	COOH	COOH

A, B, and C are all stereoisomers. A and B are enantiomers. A and C are diastereomers (epimers, too).

Special diastereomers

Cis- and trans- forms are diastereomers that result from configuration about a double bond:

CIS-: (X)(H)C=C(X)(H)

TRANS-: (H)(X)C=C(X)(H)

Anomers (designated by alpha and beta) are sugar diastereomers in which the free carbonyl (−c=o) group opens up to make a chain and two new diastereomers (see fig. 3.1).

The functional difference between alpha and beta anomers is well-illustrated in the difference between amylose and cellulose. Amylose contains 1–4 linkages between alpha-glucose molecules (fig. 3.1) whereas cellulose contains 1–4 linkages between beta glucose molecules. Humans can digest amylose but lack the enzyme to digest cellulose.

APPENDIX II. ENZYME GLOSSARY

Enzymes may be categorized as follows:

HYDROLASES
- Esterases
- Lipases
- Nucleosidases and -tidases
- Peptidases
- Phosphatases
- Sulfatases

ISOMERASES
- Epimerases
- certain Mutases
- Racemases

LIGASES
- Carboxylases
- Synthetases

LYASES
- Aldolases
- Decarboxylases
- Hydratases
- Synthases

OXIDOREDUCTASES
- Dehydrogenases
- Desaturases
- Hydroxylases
- Oxidases
- Oxygenases
- Reductases

TRANSFERASES
- Kinases
- certain Mutases
- Phosphorylases
- Polymerases
- Transaldolase and -ketolase
- Transaminases

Aldolase—a lyase that results in the formation of a carbon-carbon bond by combining an aldehyde (or ketone)(*) with another compound.
Example:

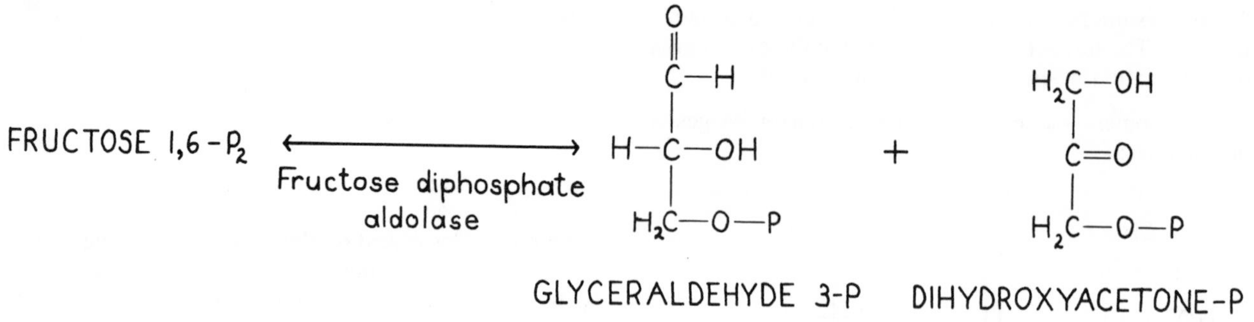

FRUCTOSE 1,6-P_2 ⟷ (Fructose diphosphate aldolase) GLYCERALDEHYDE 3-P ($O{=}CH{-}CH(OH){-}H_2C{-}O{-}P$) + DIHYDROXYACETONE-P ($H_2C(OH){-}C{=}O{-}H_2C{-}O{-}P$)

Carboxylase—a ligase that catalyzes the addition of CO_2.
Example:

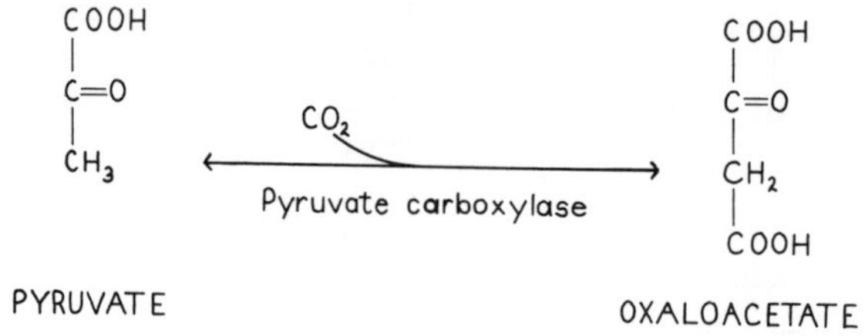

PYRUVATE ($COOH{-}C{=}O{-}CH_3$) + CO_2 ⟷ (Pyruvate carboxylase) OXALOACETATE ($COOH{-}C{=}O{-}CH_2{-}COOH$)

Decarboxylase—a lyase that catalyzes the removal of CO_2.

(*)Definitions: aldehyde, $R{-}\overset{O}{\overset{\|}{C}}{-}H$; ketone, $R{-}\overset{O}{\overset{\|}{C}}{-}R$; alcohol, $R{-}\overset{OH}{\overset{|}{C}}{-}R$ or $R{-}\overset{OH}{\overset{|}{C}}{-}H$; ester, $R{-}O{-}R$; carboxylic acid, $R{-}COOH$, where R=carbon-containing group.

Dehydrogenase—an oxidoreductase that facilitates passage of hydrogen from one molecule to another, thereby oxidizing one and reducing the other.
Example:

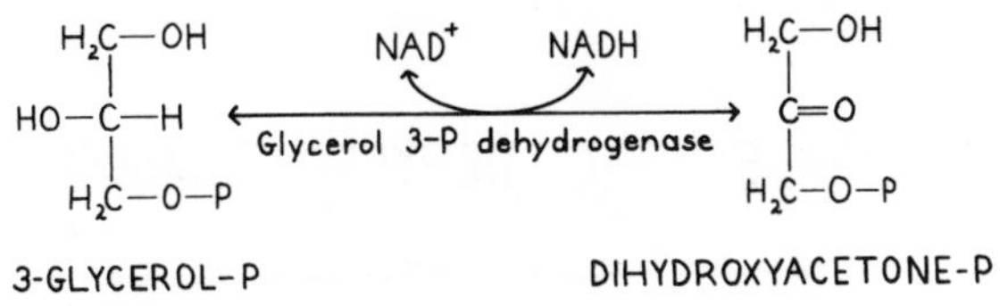

$H_2C{-}OH$ / $HO{-}C{-}H$ / $H_2C{-}O{-}P$ ⟷ (NAD⁺ → NADH; Glycerol 3-P dehydrogenase) ⟷ $H_2C{-}OH$ / $C{=}O$ / $H_2C{-}O{-}P$

NAD^+ NADH

Glycerol 3-P dehydrogenase

3-GLYCEROL-P DIHYDROXYACETONE-P

Desaturase—an oxidoreductase that catalyzes desaturation (formation of a double bond) in a fatty acid.

Epimerase—an isomerase that interconverts epimers.
Example:

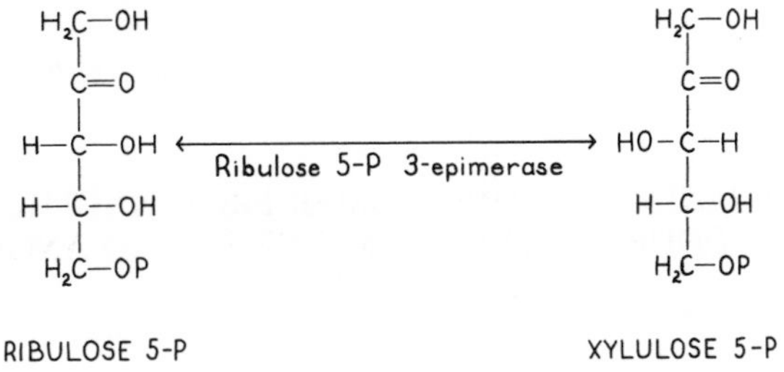

$H_2C{-}OH$ / $C{=}O$ / $H{-}C{-}OH$ / $H{-}C{-}OH$ / $H_2C{-}OP$ ⟷ $H_2C{-}OH$ / $C{=}O$ / $HO{-}C{-}H$ / $H{-}C{-}OH$ / $H_2C{-}OP$

Ribulose 5-P 3-epimerase

RIBULOSE 5-P XYLULOSE 5-P

Esterase—a hydrolase that hydrolyzes an ester into its components: an alcohol and an acid.
Example:

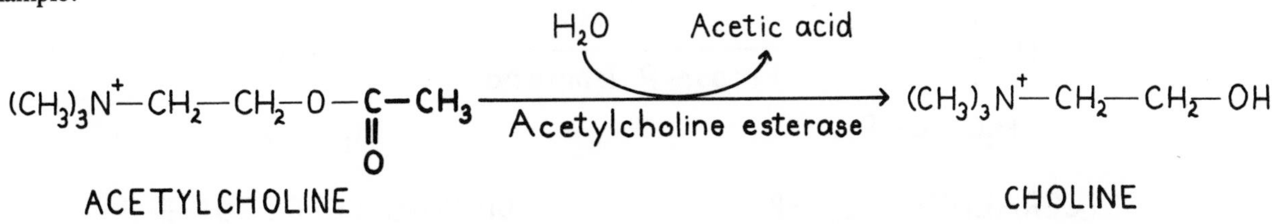

$$(CH_3)_3\overset{+}{N}{-}CH_2{-}CH_2{-}O{-}\underset{\|}{\overset{}{C}}(=O){-}CH_3 \xrightarrow[\text{Acetylcholine esterase}]{H_2O \quad \text{Acetic acid}} (CH_3)_3\overset{+}{N}{-}CH_2{-}CH_2{-}OH$$

ACETYLCHOLINE CHOLINE

Hydratase—a lyase that utilizes hydration in the formation of a C–O bond (used to be called a hydrase).
Example:

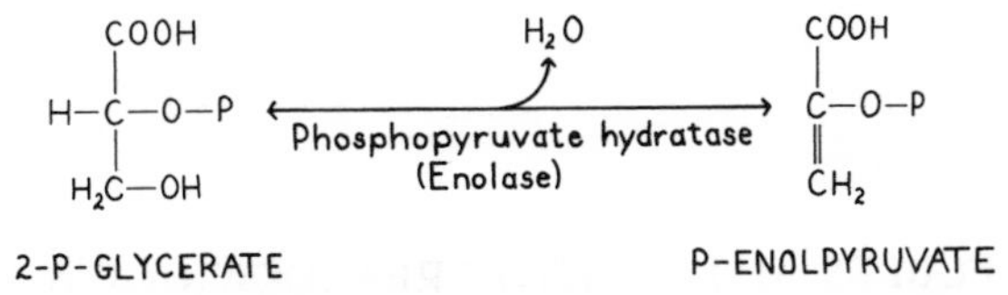

$COOH$ / $H{-}C{-}O{-}P$ / $H_2C{-}OH$ ⟷ $COOH$ / $C{-}O{-}P$ / $\|$ / CH_2 (releases H_2O)

Phosphopyruvate hydratase (Enolase)

2-P-GLYCERATE P-ENOLPYRUVATE

HYDROLASE—CLEAVES MOLECULES USING HYDROLYSIS.
Example:

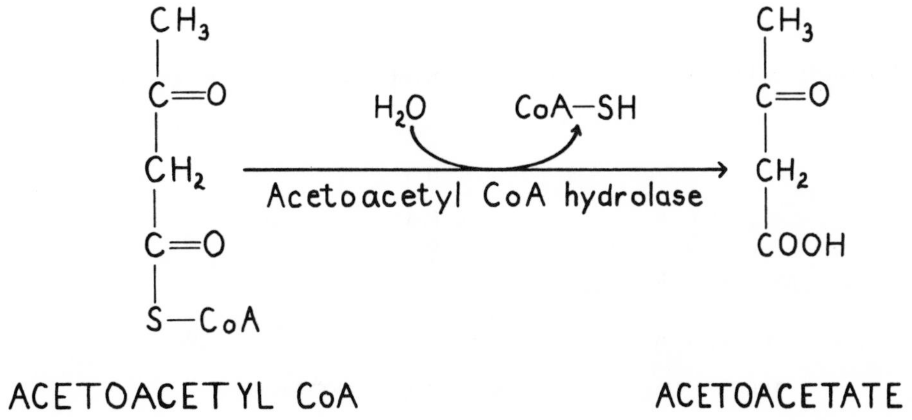

CH_3 / $C{=}O$ / CH_2 / $C{=}O$ / $S{-}CoA$ → CH_3 / $C{=}O$ / CH_2 / $COOH$

H_2O → CoA–SH

Acetoacetyl CoA hydrolase

ACETOACETYL CoA ACETOACETATE

Hydroxylase—an oxidoreductase that facilitates coupled oxidation of two donors, with incorporation of oxygen into one of the donors, oxidation of the other donor, and formation of water.
Example:

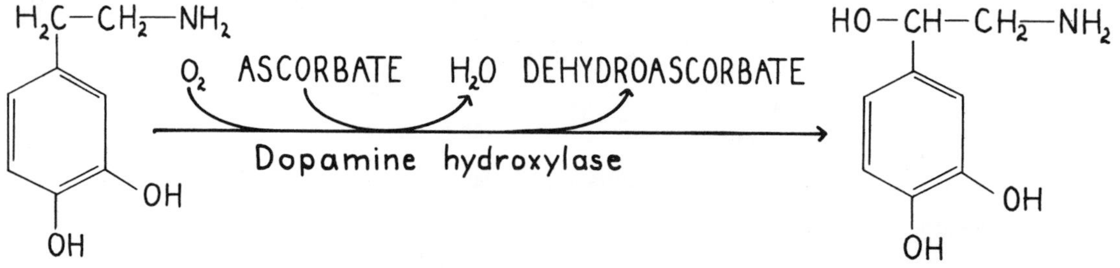

DOPAMINE $\xrightarrow[\text{Dopamine hydroxylase}]{O_2\text{, ASCORBATE} \rightarrow H_2O\text{, DEHYDROASCORBATE}}$ NOREPINEPHRINE

ISOMERASE—CATALYZES FORMATION OF ONE ISOMER FROM ANOTHER (E.G., AN ALDEHYDE TO A KETONE FORM, OR MOVING A DOUBLE BOND FROM ONE SITE TO ANOTHER ON A MOLECULE).
Example:

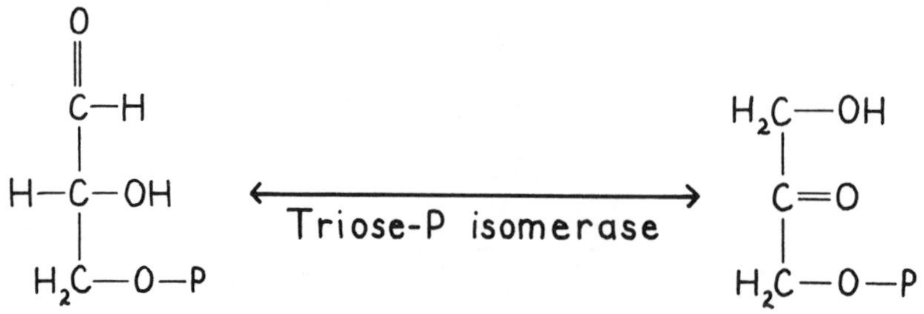

GLYCERALDEHYDE 3-P $\xleftrightarrow{\text{Triose-P isomerase}}$ DIHYDROXYACETONE-P

Kinase—a transferase that transfers a high energy group (usually P, from ATP) to an acceptor.
Example:

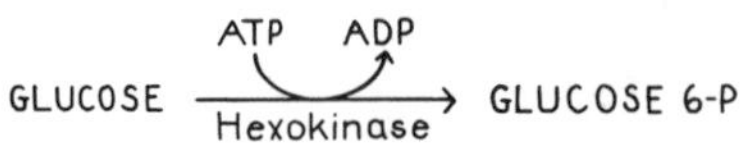

GLUCOSE $\xrightarrow[\text{Hexokinase}]{\text{ATP} \rightarrow \text{ADP}}$ GLUCOSE 6-P

LIGASE—JOINS TWO MOLECULES ALONG WITH BREAKDOWN OF A PYROPHOSPHATE (P–P) BOND. ALSO CALLED A SYNTHETASE.
Example:

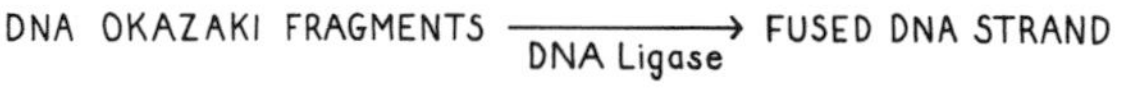

DNA OKAZAKI FRAGMENTS $\xrightarrow[\text{DNA Ligase}]{}$ FUSED DNA STRAND

Lipase—a hydrolase that catalyzes the breakup of ester linkages between a fatty acid and glycerol, within triglycerides and phospholipids.
Example:

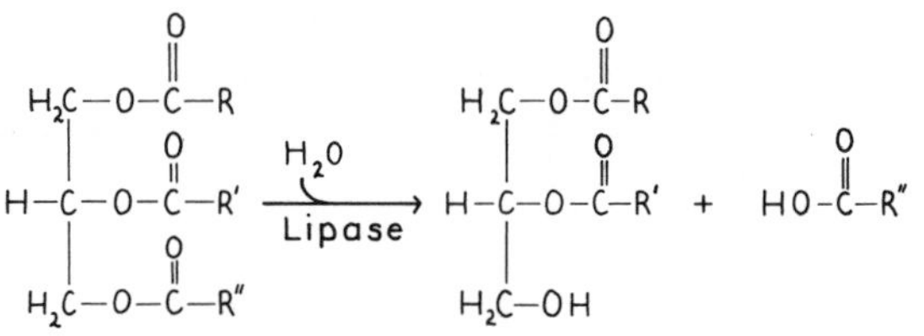

TRIGLYCERIDE $\xrightarrow[\text{Lipase}]{H_2O}$ DIGLYCERIDE + FATTY ACID

LYASE—CLEAVES OR SYNTHESIZES C–C, C–O, C–N AND OTHER BONDS BY OTHER MEANS THAN BY OXIDATION OR HYDROLYSIS (DISTINGUISHING THEM FROM OXIDOREDUCTASES AND HYDROLASES). THE REACTION FOLLOWS THE ARRANGEMENT: A + B ⟶ C (THEREBY DISTINGUISHING THEM FROM TRANSFERASES, WHICH FOLLOW THE ARRANGEMENT: A + B ⟶ C + D).
Example:

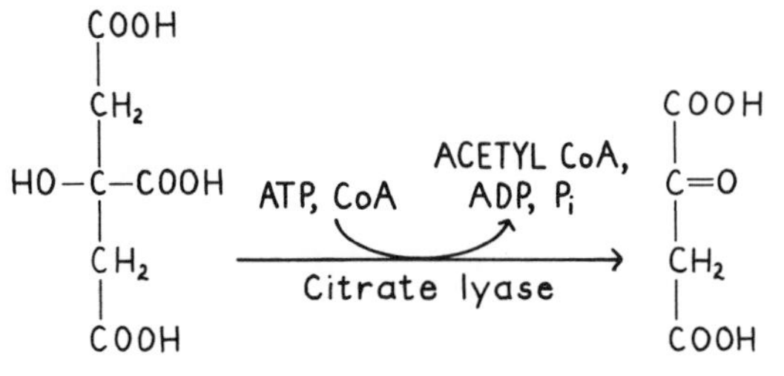

CITRATE (COOH–CH_2–C(OH)(COOH)–CH_2–COOH) + ATP, CoA ⟶ (Citrate lyase) OXALOACETATE (COOH–C=O–CH_2–COOH) + ACETYL CoA, ADP, P_i

Mutase—a transferase or isomerase that shifts one group (e.g., acyl, amino, phospho, etc.) from one point to another on the molecule.
Example:

GLUCOSE 6-P ⟶ (Phosphoglucomutase) GLUCOSE 1-P

Nucleosidase—a hydrolase that breaks nucleosides into a purine or pyrimidine and ribose.

Nucleotidase—a hydrolase that breaks down nucleotides to nucleosides and phosphoric acid. Also called phosphonuclease and nucleophosphatase.
Examples:

URIDINE 5'-PHOSPHATE ⟶ (5' nucleotidase; P_i) URIDINE ⟶ (Uridine nucleosidase; RIBOSE) URACIL

Oxidase—an oxidoreductase that catalyzes reactions in which molecular oxygen is reduced.
Example:

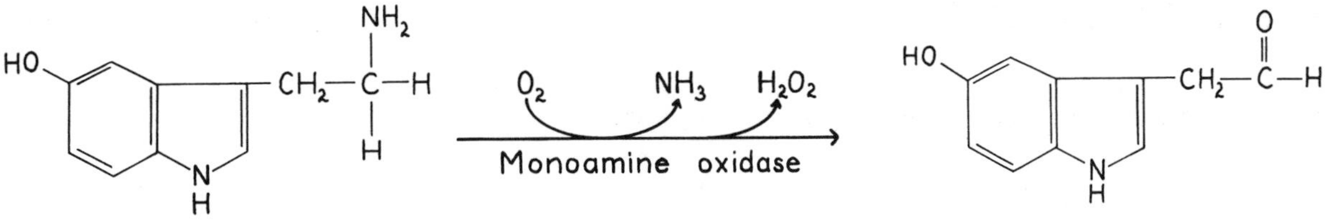

SEROTONIN (HO–indole–CH_2–$C(NH_2)H_2$) + O_2 ⟶ (Monoamine oxidase) 5-HYDROXYINDOLE ACETALDEHYDE (HO–indole–CH_2–C(=O)–H) + NH_3 + H_2O_2

OXIDOREDUCTASE—CATALYZES OXIDATION-REDUCTION REACTIONS.
Oxygenase—an oxidoreductase that catalyzes the addition of oxygen into a molecule.
Example:

HOMOGENTISATE + O_2 ⟶ (Homogentisate oxygenase (oxidase)) 4-MALEYLACETOACETATE

Peptidase—a hydrolase that facilitates the hydrolysis of a peptide bond.
Example:

$$\sim\sim N(H)-C(H)(R')-C(=O)-N(H)-C(H)(R'')-C(=O)\sim\sim \xrightarrow[\text{Peptidase}]{H_2O} \sim\sim N(H)-C(H)(R')-C(=O)-O^- + {}^+H_3N-C(H)(R'')-C(=O)\sim\sim$$

PEPTIDE

Phosphatase—a hydrolase that hydrolyzes esters, releasing inorganic phosphate.
Example:

$$\text{GLUCOSE 6-P} \xrightarrow[\text{Glucose 6-phosphatase}]{H_2O \qquad P_i} \text{GLUCOSE}$$

Phosphorylase—a transferase that catalyzes the breakage of a C–O bond by the addition of inorganic phosphate.
Example:

$$(\text{AMYLOSE})_n \xrightarrow[\text{Phosphorylase}]{P_i} \text{GLUCOSE 1-P} + (\text{AMYLOSE})_{n-1}$$

Polymerase—a transferase that catalyzes polymerization.
Example:
Formation of DNA and RNA by polymerization of purine and pyrimidine nucleotides (DNA and RNA polymerases).

Racemase—an isomerase that catalyzes the change of an optically active molecule to its opposite mirror image form.
Example:

$$\text{H}-\underset{O=C-S-CoA}{\overset{COOH}{C}}-CH_3 \xrightarrow{\text{Methylmalonyl CoA racemase}} CH_3-\underset{O=C-S-CoA}{\overset{COOH}{C}}-H$$

D-METHYLMALONYL CoA — L-METHYLMALONYL CoA

Reductase—an oxidoreductase that has reducing action; a hydrogenase.
Example:

$$HOOC-CH_2-C(OH)(CH_3)-CH_2-C(=O)-S-CoA \xrightarrow[\text{Hydroxymethylglutaryl-CoA reductase}]{\text{NADPH} \qquad \text{NADP}^+,\ \text{CoASH}} HOOC-CH_2-C(OH)(CH_3)-CH_2-H_2C-OH$$

3-HYDROXY-3-METHYLGLUTARYL CoA — MEVALONATE

Sulfatase—a hydrolase that hydrolyzes molecules, releasing sulfate.

Synthase—a lyase that catalyzes a synthesis that does **not** include breaking a pyrophosphate bond.
Example:

UDP - GLUCOSE → (Glycogen synthase; UDP released) AMYLOSE

Synthetase—a ligase that forms two molecules along with the breakdown of a pyrophosphate bond (as in breakdown of ATP).
Example:

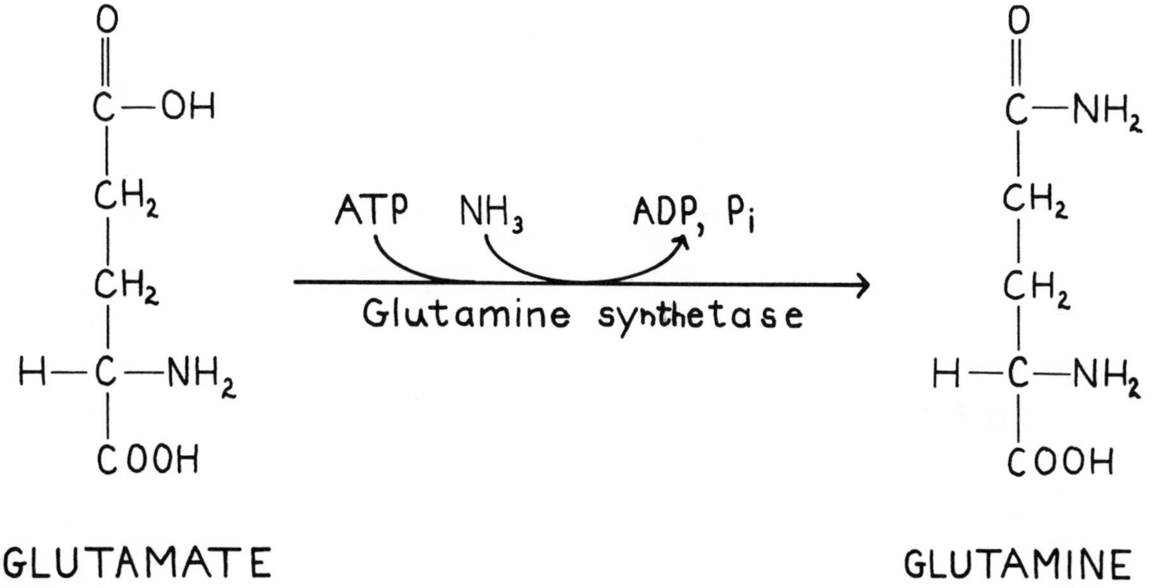

GLUTAMATE (HOOC–CH(NH$_2$)–CH$_2$–CH$_2$–C(=O)–OH) + ATP + NH$_3$ → (Glutamine synthetase) GLUTAMINE (HOOC–CH(NH$_2$)–CH$_2$–CH$_2$–C(=O)–NH$_2$) + ADP, P$_i$

Transaldolase—a transferase that transfers an aldehyde residue (e.g., in the shortening of sedoheptulose to form erythrose 4-P).
Example:

SEDOHEPTULOSE (C_7) + GLYCERALDEHYDE 3-P (C_3) → (Transaldolase) ERYTHROSE 4-P (C_4) + FRUCTOSE 6-P (C_6)

Transaminase—a transferase that transfers an amino group from an amino acid to an alpha-keto acid, especially alpha-ketoglutarate (2-ketoglutarate). Also called an aminotransferase.
Example:

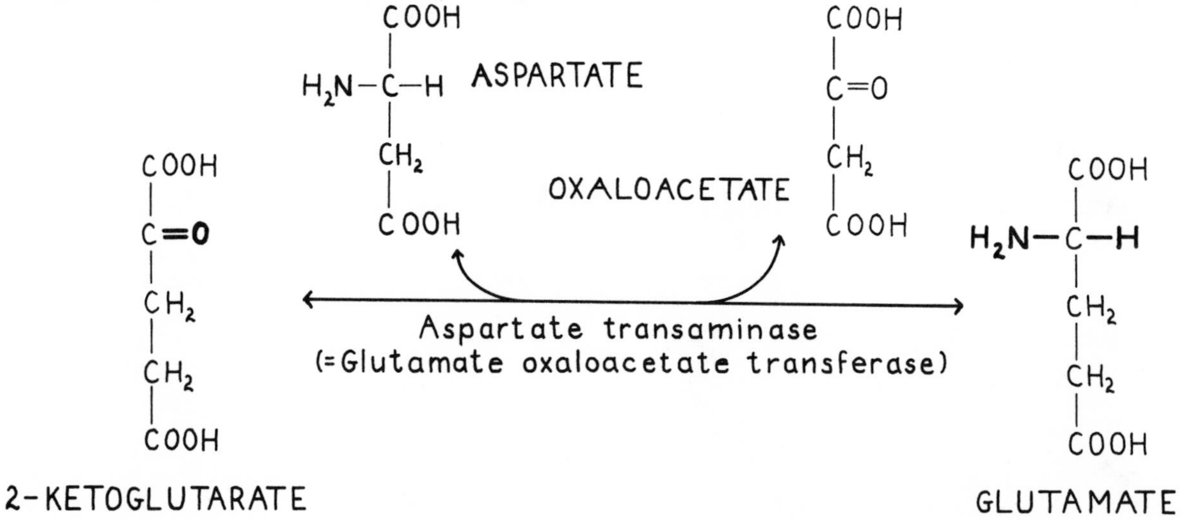

2-KETOGLUTARATE (HOOC–C(=O)–CH$_2$–CH$_2$–COOH) + ASPARTATE (HOOC–CH(NH$_2$)–CH$_2$–COOH) ⇌ (Aspartate transaminase (=Glutamate oxaloacetate transferase)) GLUTAMATE (HOOC–CH(NH$_2$)–CH$_2$–CH$_2$–COOH) + OXALOACETATE (HOOC–C(=O)–CH$_2$–COOH)

TRANSFERASE—TRANSFERS A GROUP (ASIDE FROM HYDROGEN) FROM ONE MOLECULE TO ANOTHER.
Example:

$$\text{GLUCOSE 1-P} \xrightarrow[\text{Glucose 1-P uridyl transferase}]{\text{UTP} \quad \text{PP}_i} \text{UDP-GLUCOSE}$$

Transketolase—a transferase that catalyzes the transfer of a ketone residue.
Example:

$$\begin{array}{c} \text{O} \\ \| \\ \text{C—H} \\ | \\ \text{H—C—OH} \\ | \\ \text{H}_2\text{C—O—P} \end{array} + \begin{array}{c} \text{H}_2\text{C—OH} \\ | \\ \text{C=O} \\ | \\ \text{HO—C—H} \\ | \\ \text{H—C—OH} \\ | \\ \text{H—C—OH} \\ | \\ \text{H}_2\text{C—O—P} \end{array} \xleftrightarrow[\text{Transketolase}]{} \begin{array}{c} \text{O} \\ \| \\ \text{C—H} \\ | \\ \text{H—C—OH} \\ | \\ \text{H—C—OH} \\ | \\ \text{H}_2\text{C—O—P} \end{array} + \begin{array}{c} \text{H}_2\text{C—OH} \\ | \\ \text{C=O} \\ | \\ \text{HO—C—H} \\ | \\ \text{H—C—OH} \\ | \\ \text{H}_2\text{C—O—P} \end{array}$$

GLYCERALDEHYDE 3-P FRUCTOSE 6-P ERYTHROSE 4-P XYLULOSE 5-P

STRUCTURAL INDEX

Most of the chemical structures in this index were provided courtesy of the Boehringer-Mannheim Biochemical Co.

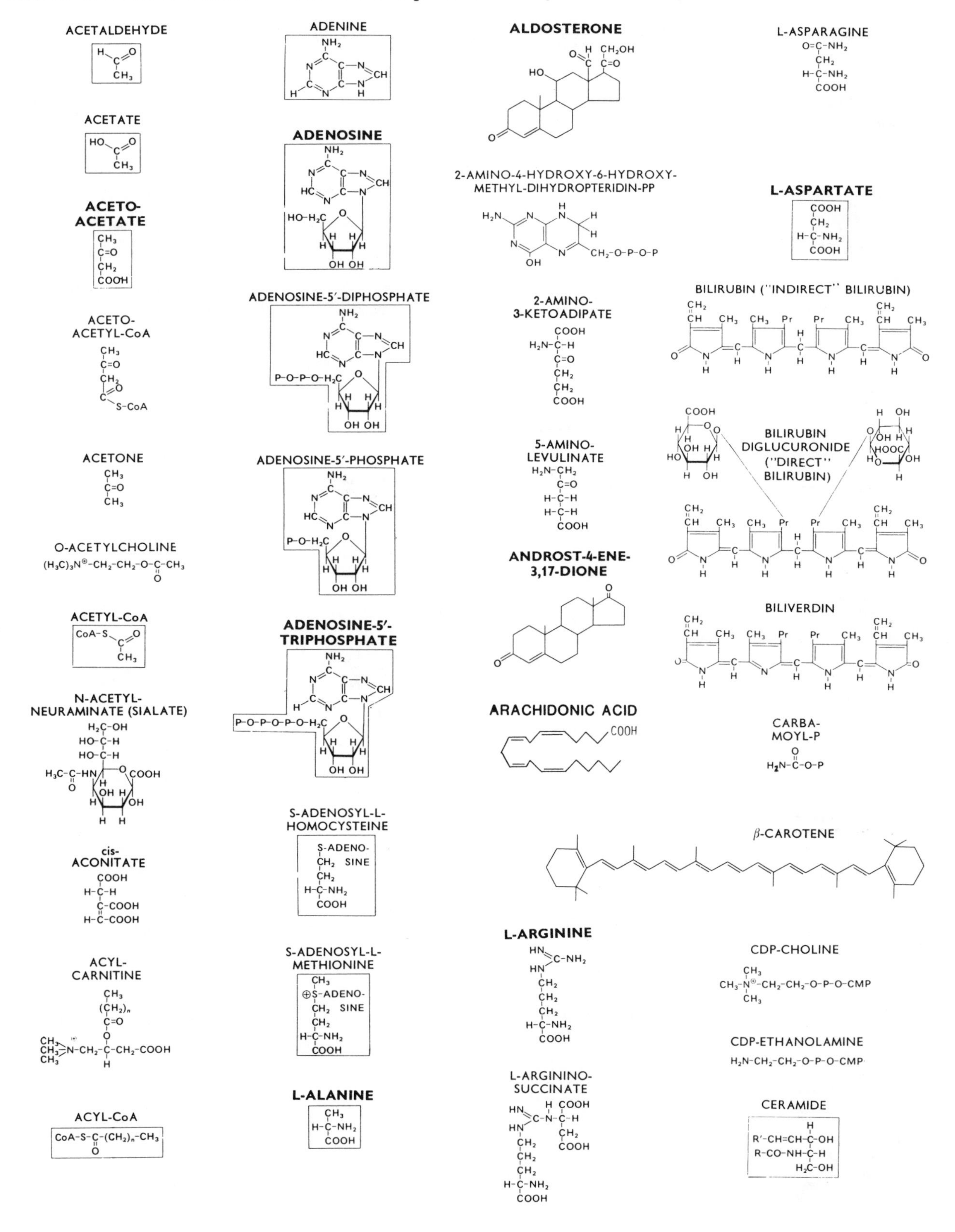

CHENODEOXYCHOLATE

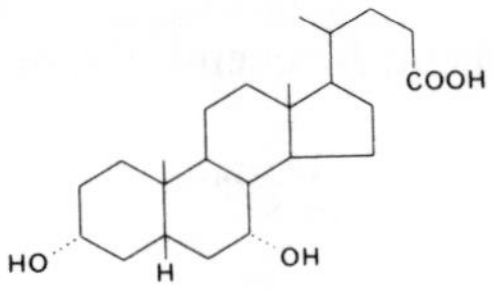

CHOLATE

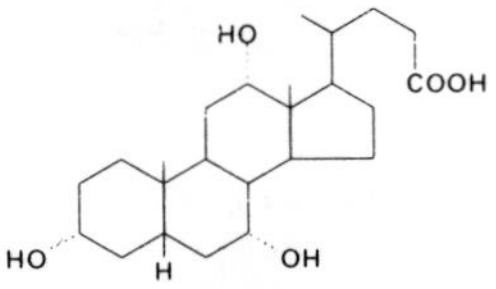

CHOLESTEROL

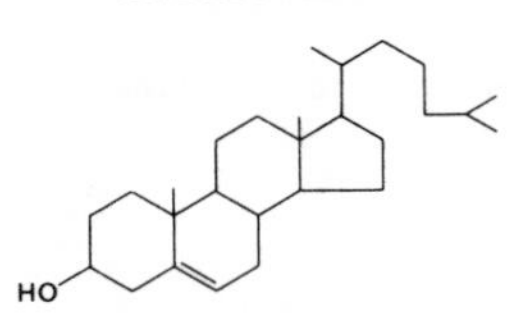

CHOLESTEROL ESTER

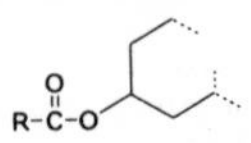

CHOLINE

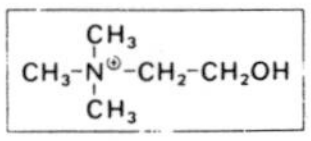

CITRATE

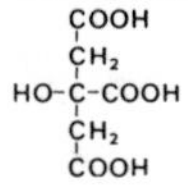

L-CITRULLINE

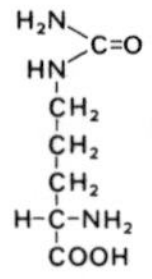

COBALAMIN (VITAMIN B_{12})

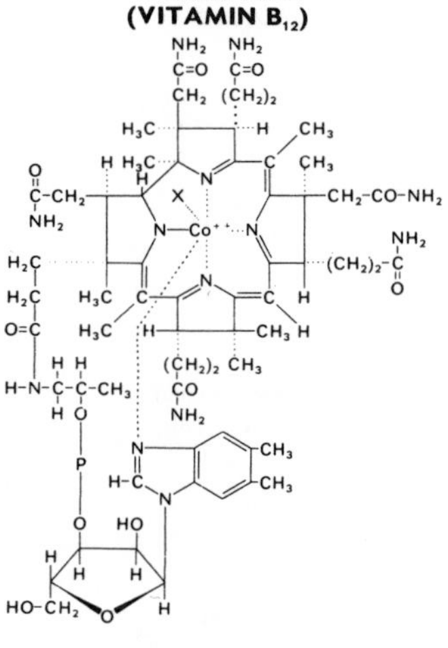

COENZYME A (CoA-SH)

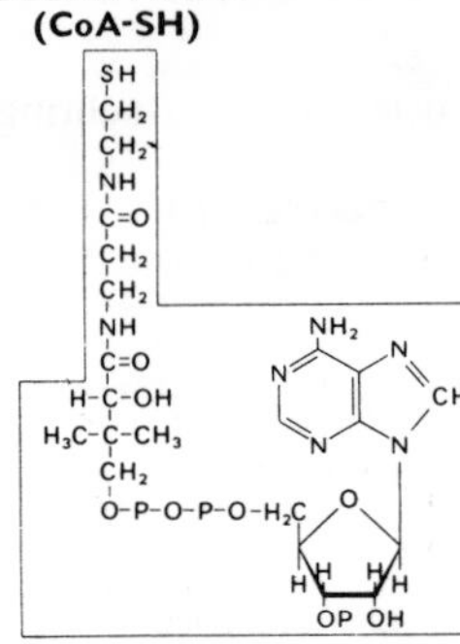

COPROPORPHYRINOGEN III

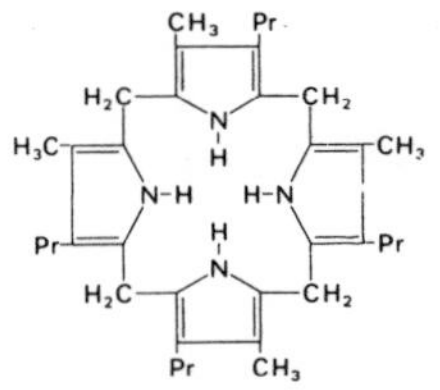

CORTICOSTERONE

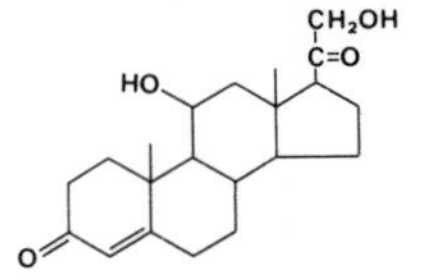

CORTISOL

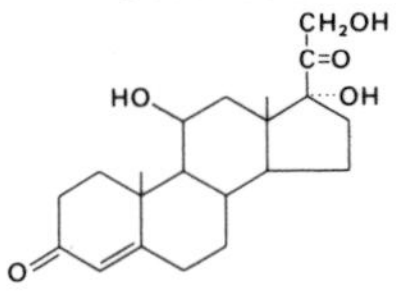

CREATINE

CREATINE-P

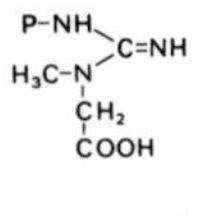

CREATININE

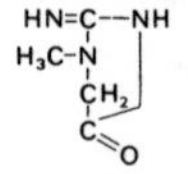

ADENOSINE-3′,5′-MONOPHOSPHATE (CYCLIC)

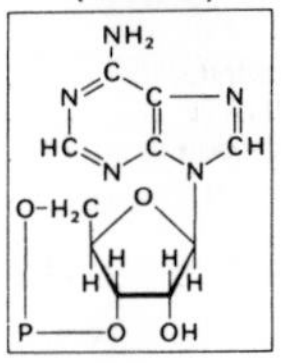

L-CYSTATHIONINE

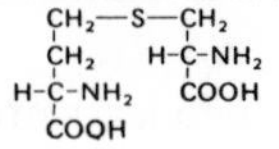

L-CYSTEINE

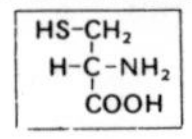

L-CYSTINE

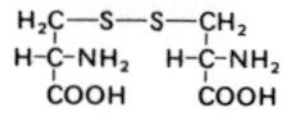

CYTIDINE

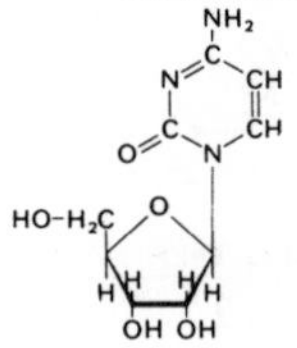

CYTIDINE-5′-DIPHOSPHATE

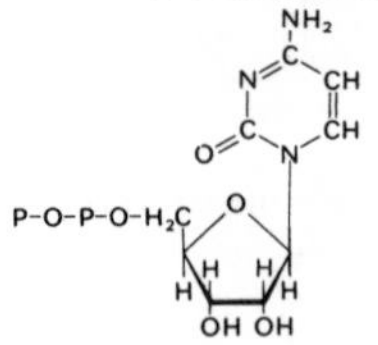

CYTIDINE-5′-PHOSPHATE

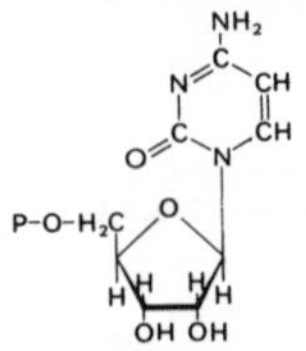

CYTIDINE-5′-TRIPHOSPHATE

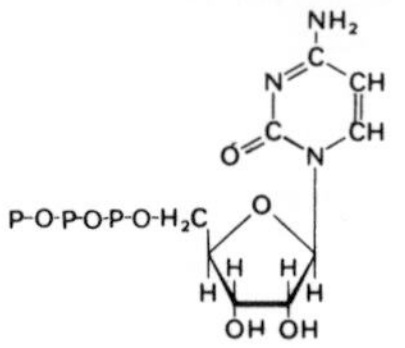

CYTOCHROME c

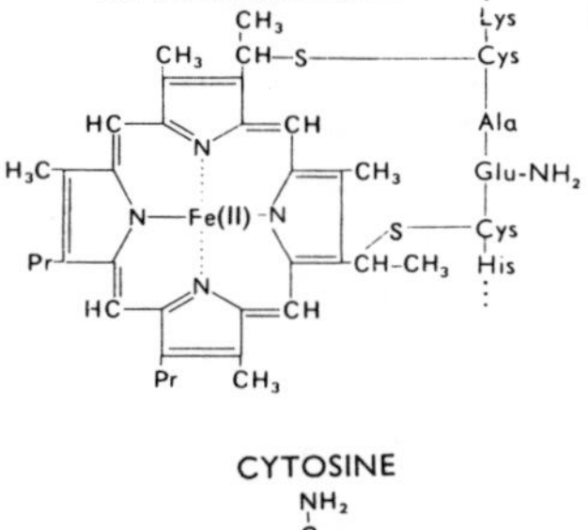

CYTOSINE

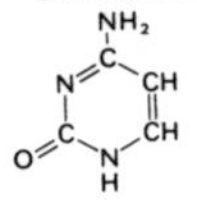

DEHYDROEPIANDROSTERONE

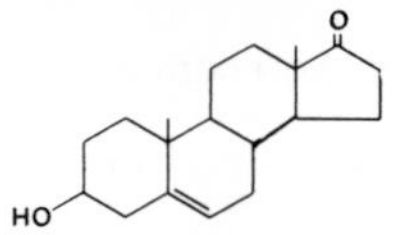

2′-DEOXYADENOSINE-5′-TRIPHOSPHATE

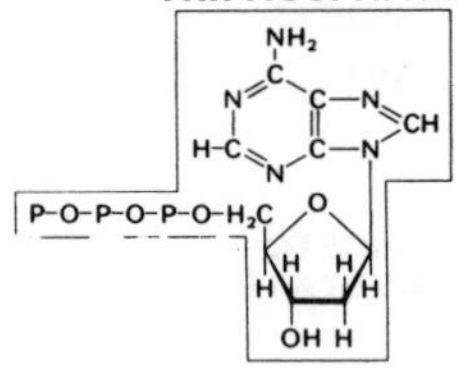

DEOXYCHOLATE

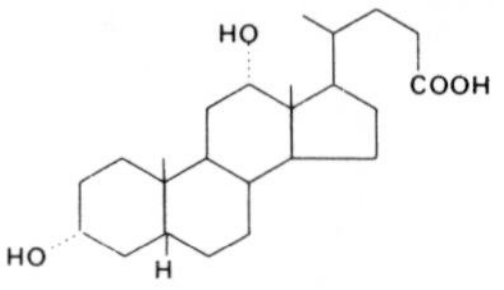

2′-DEOXYCYTIDINE-5′-DIPHOSPHATE

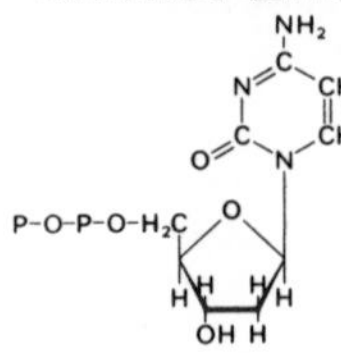

2′-DEOXYCYTIDINE-5-PHOSPHATE

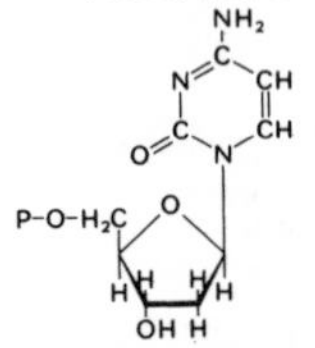

2′-DEOXYCYTIDINE-5′-TRIPHOSPHATE

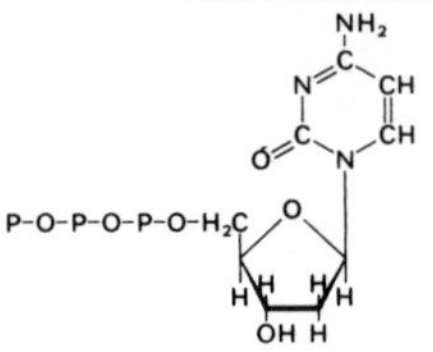

2′-DEOXYGUANOSINE-5′-DIPHOSPHATE

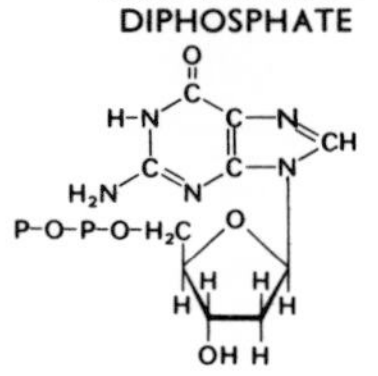

2′-DEOXY-GUANOSINE-5′-TRIPHOSPHATE

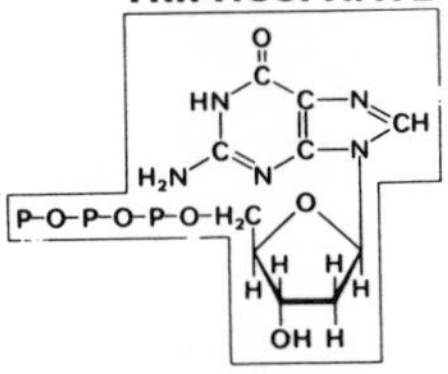

2′-DEOXY-URIDINE-5′-DIPHOSPHATE

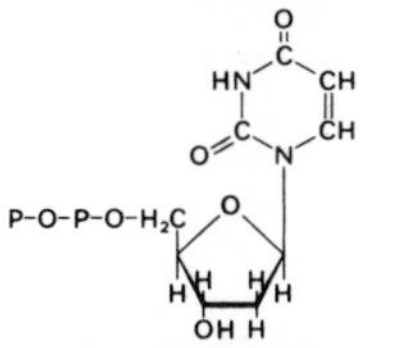

2′-DEOXY-URIDINE-5′-PHOSPHATE

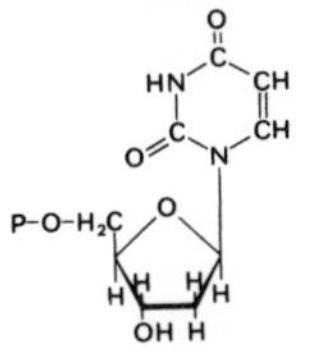

2′-DEOXY-URIDINE-5′-TRIPHOSPHATE

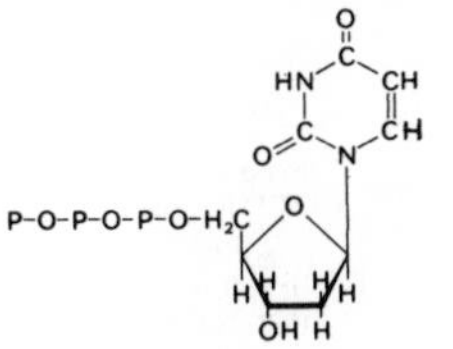

DIGLYCERIDE

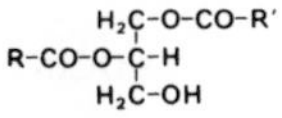

H₂C-O-CO-R′
R-CO-O-C-H
H₂C-OH

7,8-DIHYDRO-FOLATE

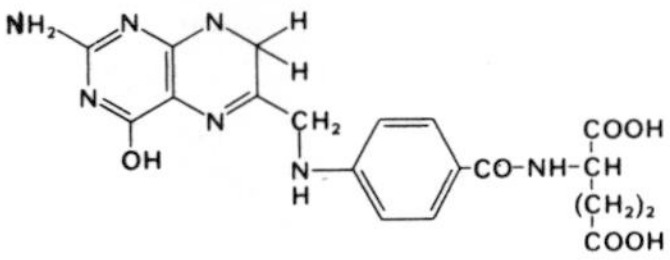

7,8-DIHYDROPTEROATE

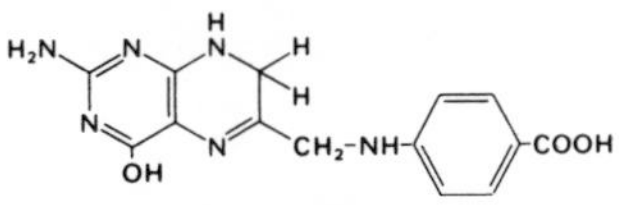

DIHYDROXY-ACETONE P

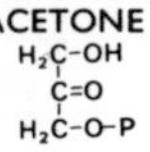

H₂C-OH
C=O
H₂C-O-P

L-DIHYDROXY-PHENYLALANINE

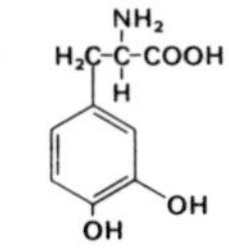

DIIODO-L-TYROSINE

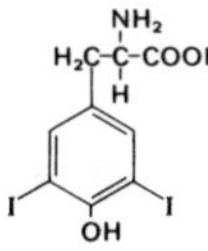

DOPAMINE

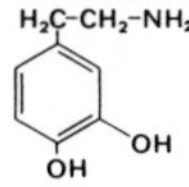

L-EPINEPHRINE

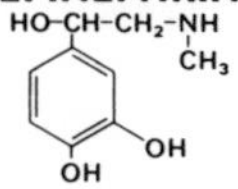

ERGOCALCIFEROL (VITAMIN D_2)

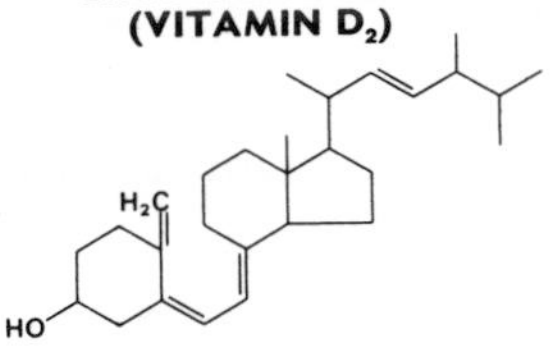

D-ERYTHROSE 4-P

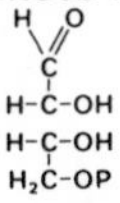

ESTRADIOL-17β

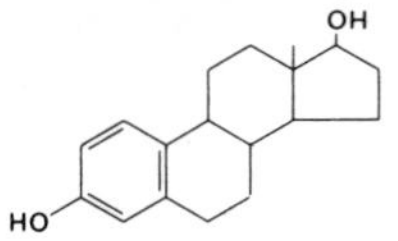

ESTRIOL

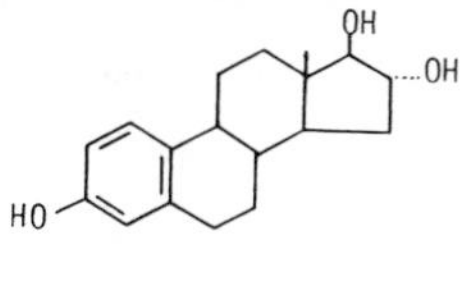

ESTRONE

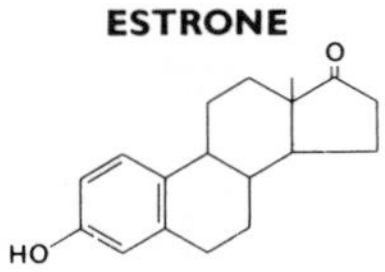

ETHANOL

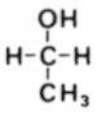

FATTY ACID

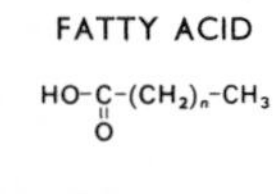

FAD (FLAVIN ADENINE DINUCLEOTIDE)

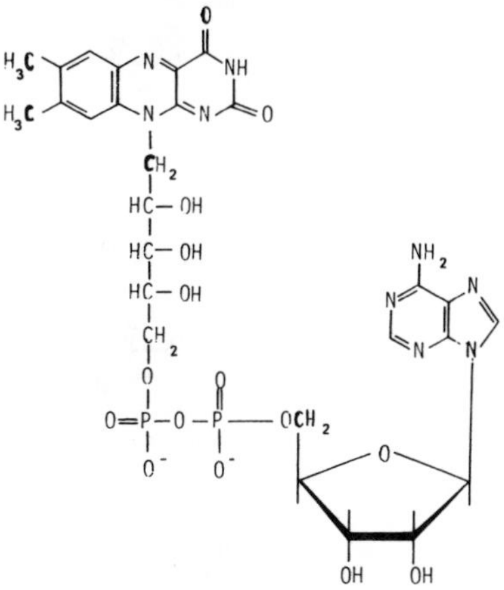

N-FORMIMINO-L-GLUTAMATE

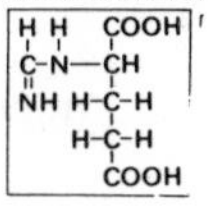

D-FRUCTOSE

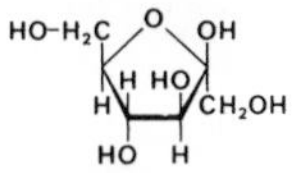

β-D-FRUCTOSE 1,6-P_2

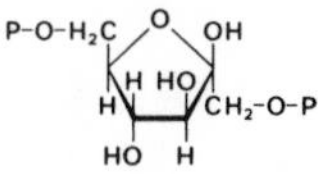

β-D-FRUCTOSE 1-P

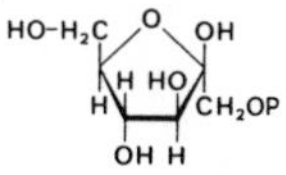

β-D-FRUCTOSE 6-P

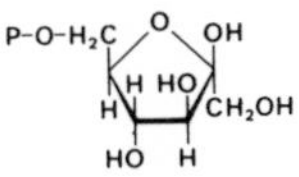

FUMARATE

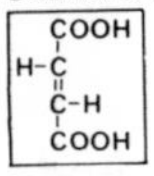

GABA

4-AMINO-BUTYRATE

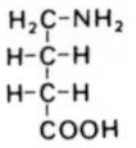

D-GALACTOSE

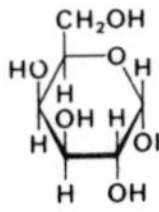

α-D-GALACTOSE 1-P

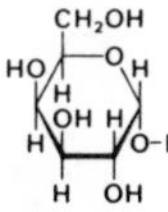

D-6-P-GLUCONATE

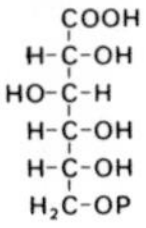

D-6-P-GLUCONO-δ-LACTONE

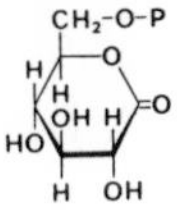

D-GLUCOSAMINE 6-P

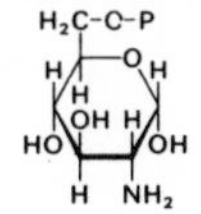

D-GLUCOSE

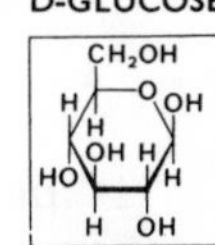

α-D-GLUCOSE 1-P

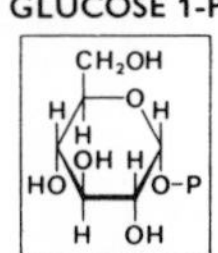

α-D-GLUCOSE 6-P

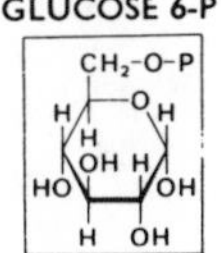

L-GLUTAMATE

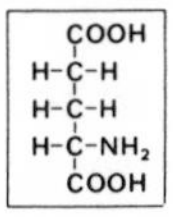

L-GLUTAMINE

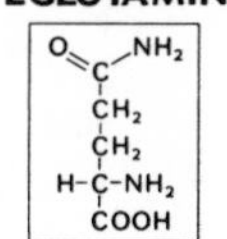

RED. GLUTATHIONE (GSH)

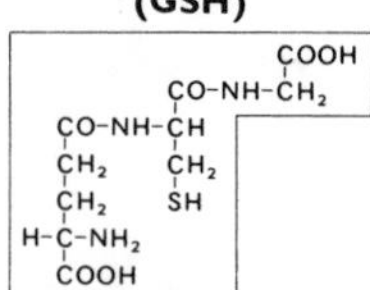

D-GLYCERALDEHYDE

D-GLYCERALDEHYDE 3-P

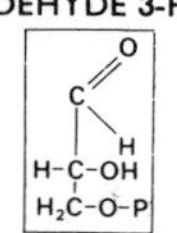

1,3-P_2-D-GLYCERATE

2,3-P_2-D-GLYCERATE

2-P-D-GLYCERATE

3-P-D-GLYCERATE

GLYCEROL

L-3-GLYCEROL P

GLYCINE

GLYCOCHOLATE

GUANINE

GUANOSINE

GUANOSINE-5 -DIPHOSPHATE

GUANOSINE-5′-PHOSPHATE

GUANOSINE-5′-TRIPHOSPHATE

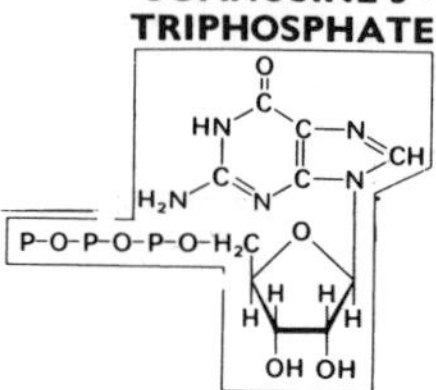

HEME

HEMOGLOBIN

HISTAMINE

L-HISTIDINE

L-HOMOCYSTEINE

HOMOGENTISATE

D-3-HYDROXYBUTYRATE

5-HYDROXYINDOLE ACETALDEHYDE

5-HYDROXYINDOLE ACETATE

3-HYDROXY-L-KYNURENINE

3-HYDROXY-3-METHYLGLUTARYL-CoA

p-HYDROXYPHENYLPYRUVATE

17α-HYDROXYPREGNENOLONE

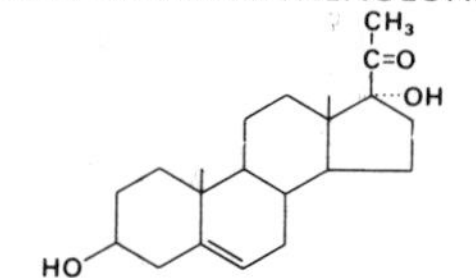

17α-HYDROXYPROGESTERONE

L-4-HYDROXYPROLINE

HYPOXANTHINE

INOSINE

INOSINE-5′-PHOSPHATE

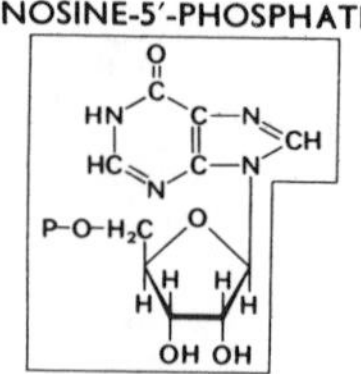

threo-Ds-ISOCITRATE

L-ISOLEUCINE

Δ^3ISOPENTENYL-PP

2-KETOBUTYRATE

2-KETOGLUTARATE

STRUCTURAL INDEX

L-LACTATE

COOH
HO-C-H
CH_3

LACTOSE

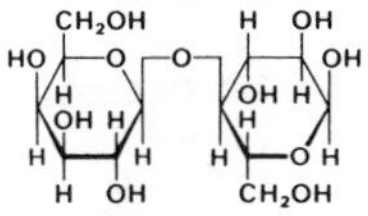

CH_2OH H OH
HO O O OH
H OH H OH H
OH H H H O H
H H H H
H OH CH_2OH

L-LEUCINE

CH_3
H_3C-C-H
CH_2
H-C-NH_2
COOH

L-LYSINE

H_2C-NH_2
$(CH_2)_3$
H-C-NH_2
COOH

L-MALATE

COOH
HO-C-H
CH_2
COOH

4-MALEYL-ACETOACETATE

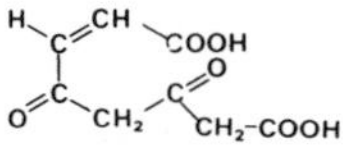

H-C=CH COOH
O=C-CH_2-C=O CH_2-COOH

MALONYL-CoA

O=C-CH_2-COOH
CoA-S

MALTOSE

Glucosyl-α(1 → 4)-Glucose

L-METHIONINE

CH_2-S-CH_3
CH_2
H-C-NH_2
COOH

R(=L)-METHYLMALONYL-CoA

COOH
H_3C-C-H
O=C-S-CoA

MEVALONATE

COOH
CH_2
H_3C-C-OH
CH_2
H_2C-OH

MONO-GLYCERIDE

H_2C-O-CO-R
HO-C-H
H_2C-OH

MONOIODO-L-TYROSINE

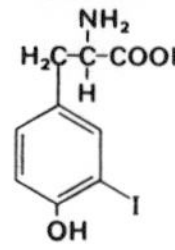

NH_2
H_2C-C-COOH
H
OH I

NIACIN

(NICOTINATE)

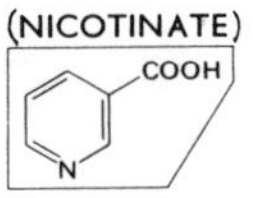

COOH
N

NICOTINAMIDE

O
C-NH_2
N

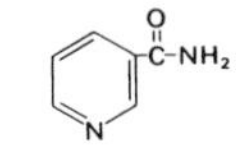

NICOTINAMIDE ADENINE DINUCLEOTIDE, NAD (DPN)

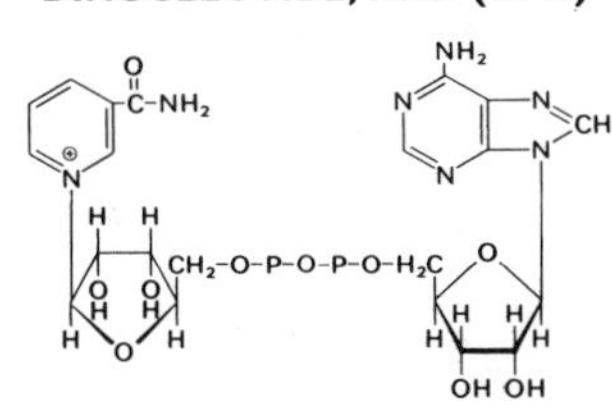

L-NOR-EPINEPHRINE

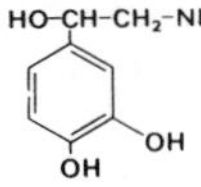

HO-CH-CH_2-NH_2
OH
OH

L-ORNITHINE

NH_2
CH_2
CH_2
CH_2
H-C-NH_2
COOH

OROTATE

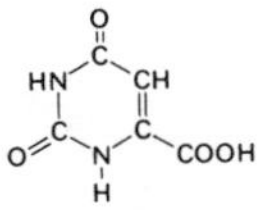

O
HN C CH
O=C N C COOH
H

OXALOACETATE

COOH
C=O
CH_2
COOH

PALMITOYL-CoA

H_3C-$(CH_2)_{14}$-C=O S-CoA

L-PHENYLALANINE

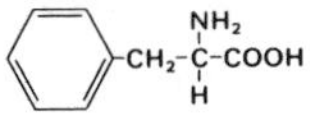

NH_2
CH_2-C-COOH
H

PHENYL-PYRUVATE

O
CH_2-C-COOH

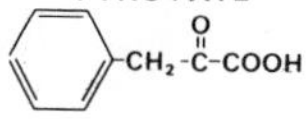

P_i (INORGANIC PHOSPHATE)

HPO_4^{-2}

PHOSPHATIDYL CHOLINE (L-1-LECITHIN)

H_2C—O—CO—R'
R—CO—O—C—H
H_2C—O—P—O—$(CH_2)_2$—$N^+(CH_3)_3$

L-1-PHOSPHATIDYL-ETHANOLAMINE

H_2C-O-CO-R'
R-CO-O-C-H
H_2C-O-P-O-$(CH_2)_2$-NH_2

L-1-PHOSPHATIDYL-INOSITOL

H_2C-O-CO-R'
R-CO-O-C-H
H_2C-O-P-O-INOSITOL

L-1-PHOSPHATIDYL-SERINE

H_2C-O-CO-R'
R-CO-O-C-H NH_2
H_2C-O-P-O-CH_2-C-COOH
H

P-ENOL-PYRUVATE

COOH
C-O-P
CH_2

5-P-β-D-RIBOSYL-AMINE

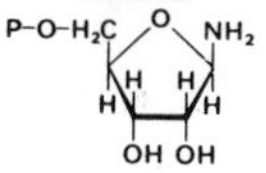

P-O-H_2C O NH_2
H H H H
OH OH

α-D-5-P-RIBOSYL-PP (PRPP)

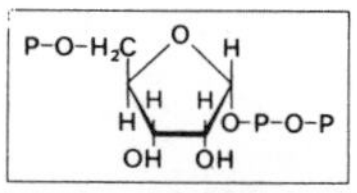

P-O-H_2C O H
H H H O-P-O-P
OH OH

PORPHOBILINOGEN

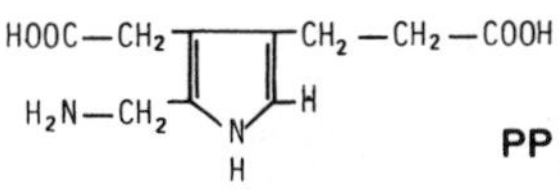

HOOC—CH_2 CH_2—CH_2—COOH
H_2N—CH_2 N H
H

PREGNENOLONE

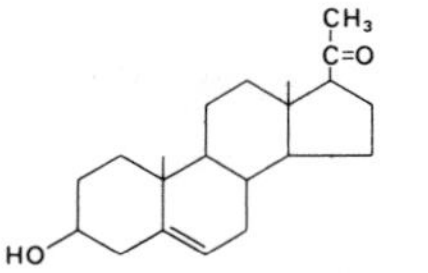

CH_3
C=O
HO

PROGESTERONE

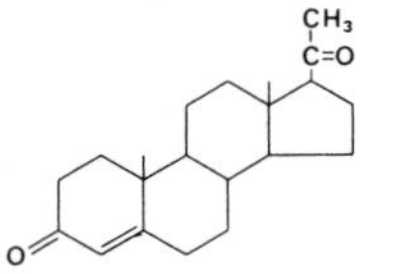

CH_3
C=O
O

L-PROLINE

H
N COOH
H

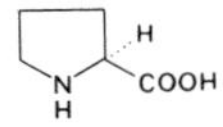

PROPIONATE

CH_3
H-C-H
O=C-OH

PROPIONYL-CoA

CH_3
H-C-H
O=C-S-CoA

PROSTAGLANDINS (PGE_1 shown)

HO H HO
O H COOH

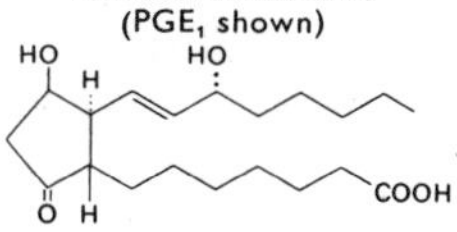

PROTOPORPHYRIN IX

CH_2
CH_3 CH
HC N CH
H_3C CH_3
N-H H-N
Pr CH=CH_2
HC N CH
Pr CH_3

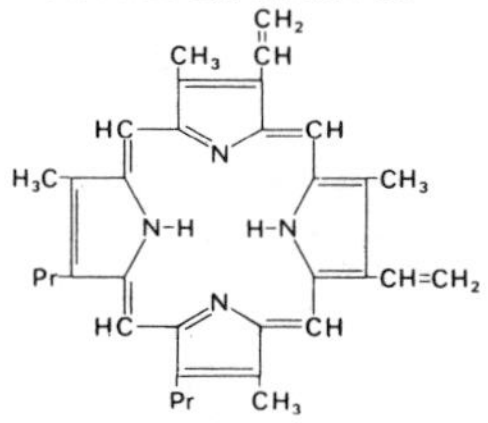

PROTO-PORPHYRINOGEN IX

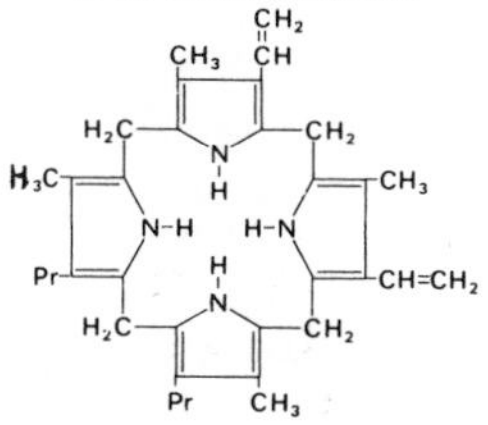

CH_2
CH_3 CH
H_2C N H CH_2
H_3C CH_3
N-H H-N
Pr CH=CH_2
H
H_2C N CH_2
Pr CH_3

PYRIDOXAL PHOSPHATE

O O=C—H
O^-—P—OH_2C OH
O^- N^+ CH_3
H

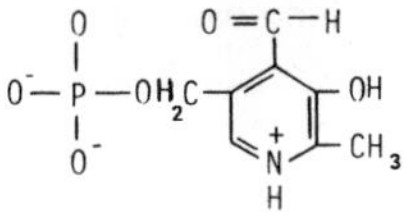

PP_i (PYROPHOSPHORIC ACID)

O O
HO—P—O—P—OH
OH OH

PYRUVATE

COOH
C=O
CH_3

RETINAL(CIS)

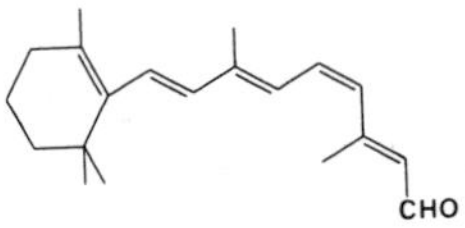

CHO

RETINAL(TRANS)

CHO

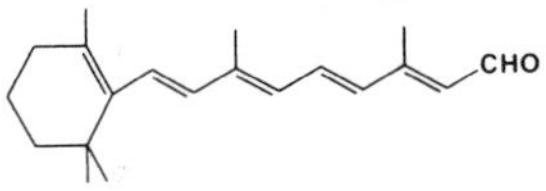

RETINOL (VITAMIN A)

D-RIBOSE

D-RIBOSE 5-P

D-RIBULOSE 5-P

D-SEDOHEPTULOSE 7-P

L-SERINE

SEROTONIN

D-SORBITOL

SPHINGOMYELIN

SPHINGOSINE

SQUALENE

I-STERCOBILIN

I-STERCOBILINOGEN

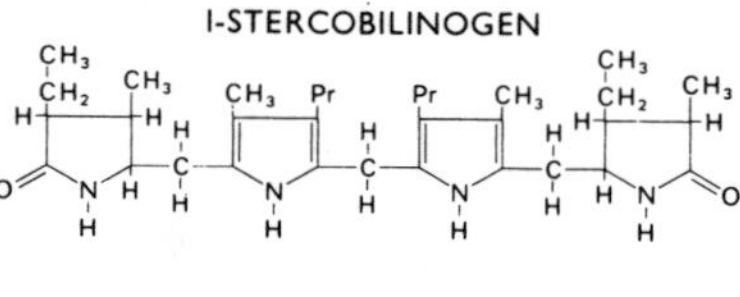

SUCCINATE

SUCCINYL-CoA

TAUROCHOLATE

TESTOSTERONE

5,6,7,8-TETRAHYDROFOLATE (THF)

THIAMINE PYROPHOSPHATE

L-THREONINE

THYMIDINE

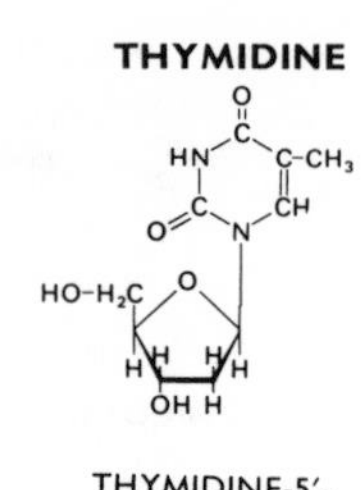

THYMIDINE-5'-DIPHOSPHATE

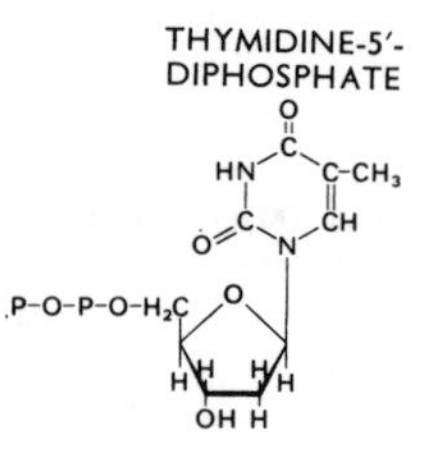

THYMIDINE-5'-PHOSPHATE

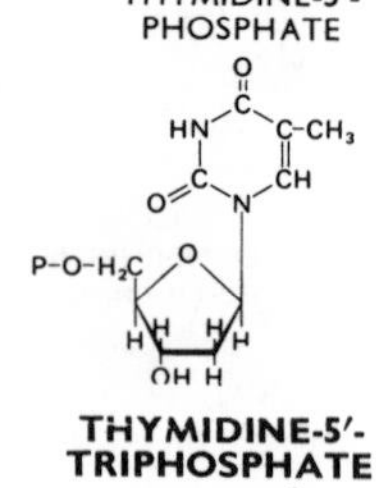

THYMIDINE-5'-TRIPHOSPHATE

THYMINE

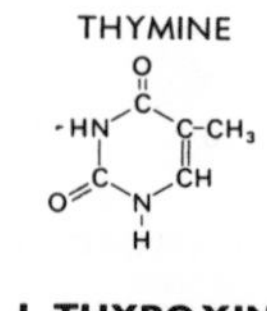

L-THYROXINE

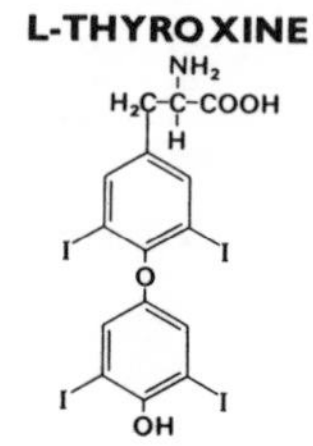

TRIGLYCERIDE

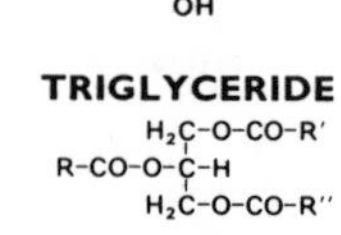

L-TRYPTOPHAN

L-TYROSINE

UDP-D-GALACTOSE

UDP-D-GLUCOSE

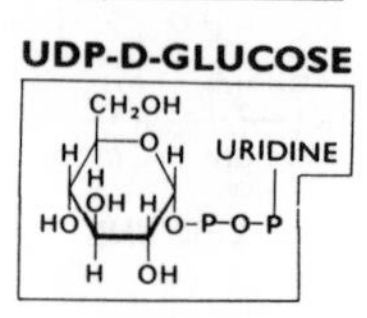

UDP-D-GLUCURONATE

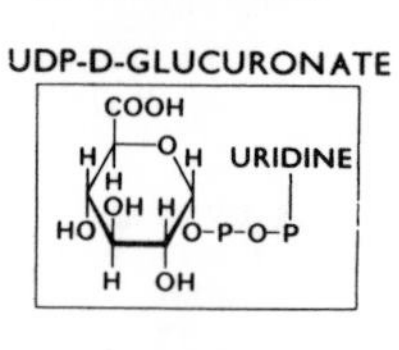

URACIL

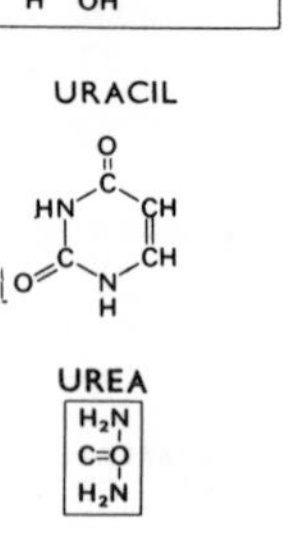

UREA

URIC ACID

URIDINE

URIDINE-5'-DIPHOSPHATE

URIDINE-5'-PHOSPHATE

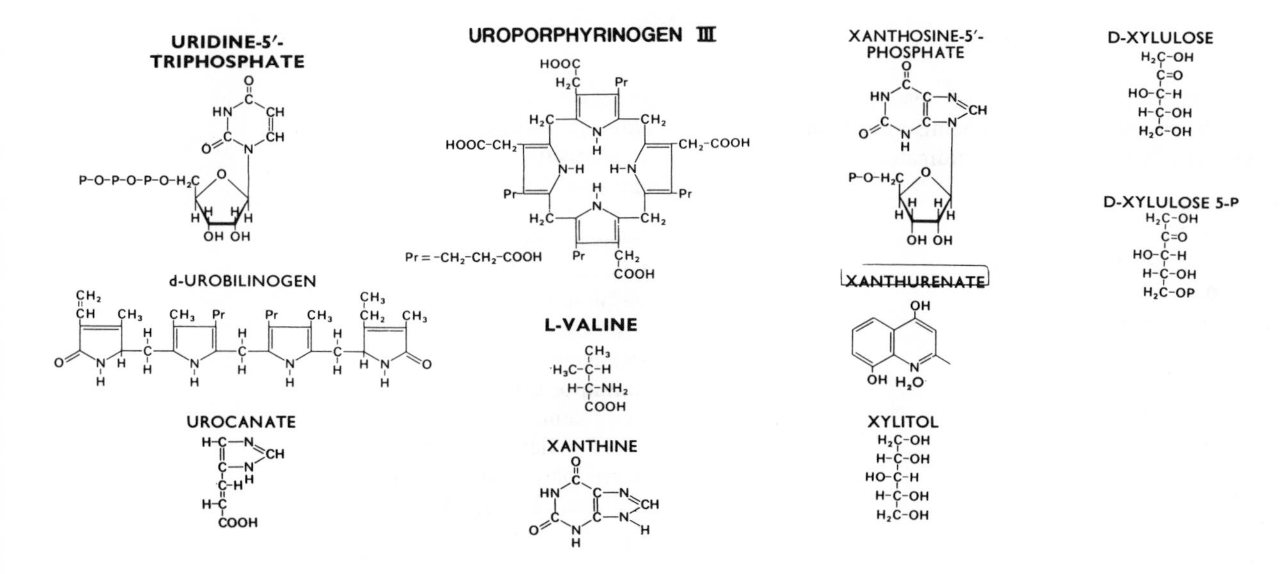

GLOSSARY

acyl: containing a fatty acid attachment

alcohol: $\text{R-}\overset{\text{OH}}{\text{C}}\text{-R}$ or $\text{R-}\overset{\text{OH}}{\text{C}}\text{-H}$

aldehyde: $\text{R-}\overset{\text{O}}{\overset{\|}{\text{C}}}\text{-H}$

anabolic: involving synthesis of more complex substances from simpler ones

carboxylic acid: R-COOH

catabolic: involving breakdown of more complex substances to simpler ones

complement: a special group of serum proteins that are important in antigen-antibody interactions

ester: R-O-R

hemolysis: red cell destruction

hydrolysis: splitting of a molecule with water, the -OH group of water becoming part of one molecule, and the hydrogen atom becoming part of the other.

ketone: $\text{R-}\overset{\text{O}}{\overset{\|}{\text{C}}}\text{-R}$

lipolysis: lipid breakdown

megaloblastic anemia: a red blood cell deficiency in which there are large red blood cells

microsomes: vesicular fragments of endoplasmic reticulum seen after intentional disruption of cells

oxidation: loss of hydrogen atoms or electrons

proteolysis: protein breakdown

pyrrol: a cyclic ring that, in groups of four, forms the basic structural unit of the porphyrins

R-: carbon-containing

reduction: addition of hydrogen or gain of electrons

substrate: the substance on which an enzyme acts (as opposed to the *product*)

INDEX

Note: Numbers preceded by a (#) refer to disease numbers on the bottom of the Biochemistryland map. Numbers in brackets refer to map coordinates.

RAPID LEARNING AND RETENTION THROUGH THE MEDMASTER SERIES:

CLINICAL NEUROANATOMY MADE RIDICULOUSLY SIMPLE, by S. Goldberg, M.D. (1990) (Spanish translation at same price). Presents the most relevant points in clinical neuroanatomy with mnemonics, humor, and case presentations. For neuroanatomy courses and Board review. 87 pgs., 61 illustr.; $10.95.

CLINICAL ANATOMY MADE RIDICULOUSLY SIMPLE, by S. Goldberg, M.D. (1990). A systemic approach to clinical anatomy, utilizing a high picture-to-text ratio. Memory is facilitated by conceptual diagrams, ridiculous associations, and a strong focus on clinical relevance. Excellent Board review; 175 pgs., 303 illustr.; $16.95.

CLINICAL ANATOMY AND PATHOPHYSIOLOGY FOR THE HEALTH PROFESSIONAL, by J.V. Stewart, M.D. (1989). For nursing students and other health professionals. Focuses on clinically relevant systemic anatomy and physiology. Brief, up-to-date, practical, thorough, enjoyable reading. 260 pgs., 230 illustr.; $16.95.

OPHTHALMOLOGY MADE RIDICULOUSLY SIMPLE, by S. Goldberg, M.D. (1990). All the ophthalmology necessary for the nonophthalmologist; 82 pgs., 75 illustr.; $10.95.

PSYCHIATRY MADE RIDICULOUSLY SIMPLE, by W.V. Good, M.D. and J. Nelson, M.D. (1991). A delightful and practical guide to clinical psychiatry; 83 pgs., 20 illustr.; $10.95.

THE 4-MINUTE NEUROLOGIC EXAM, by S. Goldberg, M.D. (1988). A rapid approach to neurologic evaluation when time is limited; 56 pgs., 13 illustr.; $8.95.

BEHAVIORAL SCIENCE FOR THE BOREDS, by F.S. Sierles, M.D. (1989). A concise review of Behavioral Science for Part I of the National Boards. Biostatistics, Medical Sociology, Psychopathology, etc. in only 133 pgs.; $12.95.

ACUTE RENAL INSUFFICIENCY MADE RIDICULOUSLY SIMPLE, by C. Rotellar, M.D. (1987). A brief, practical, and humorous approach to acute renal insufficiency. 56 pgs., 49 illustr.; $8.95.

CLINICAL BIOCHEMISTRY MADE RIDICULOUSLY SIMPLE, by S. Goldberg, M.D. (1991). A conceptual approach to clinical biochemistry, with humor. Includes a color MAP OF BIOCHEMISTRYLAND (an amusement park in which clinical biochemistry is seen as a whole, along with its key pathways, diseases, drugs, and laboratory tests). For biochemistry courses and Medical Board review. Hardcover; $20.95. (Extra maps $4.50).

JONAH: THE ANATOMY OF THE SOUL, by S. Goldberg, M.D. (1990). A new appraisal of the mind-body problem and its relations to quantum physics. Presents the strongest case to date for the presence of consciousness in computers and the persistence of the conscious mind after death; 95 pgs.; $8.95.

CLINICAL PSYCHOPHARMACOLOGY MADE RIDICULOUSLY SIMPLE, by J. Preston, Psy. D. and J. Johnson, M.D. (1990). A brief, practical review of the indications for and use of pharmacologic agents in the treatment of psychologic disorders. 42 pgs.; $8.95.

Try your bookstore for these, or, if unavailable, send the above amounts (plus $2.50 postage and handling per total order) to:

MedMaster, Inc.
P.O. Box 640028
Dept. BC
Miami, FL 33164